# EXOT[illegible] PET HANDBOOK

## David Manning

David Manning has spent his life keeping and working with animals, and especially herptofauna. He started out working for a company which supplied zoos, universities and the pet trade with specialist advice, equipment and livestock. Increasing interest from the public in more unusual animals prompted David to help to establish and manage 'The Vivarium', London's leading herpetological centre, in the 1980s.

David's own company, Animal Ark, www.animal-ark.co.uk, co-ordinate and advise film makers and photographers on using animals. David's creatures appear in television commercials and films throughout the world, and David appears with them on children's television programmes talking about animals and conservation. Animal Ark's educational roadshows also visit thousands of school children every year promoting, amongst other wildlife issues, responsible pet keeping.

## Author's Acknowledgements

To my family, especially Jenny for her invaluable assistance, and to my children Jack and Georgia, whom I hope I have not neglected during the writing of this book. Thanks also to the following: to Simon Murrell for his patience and excellent photography; to Robert Baltrock of Bio-Pet Ltd who gave me, after much persuasion, my first chance to work with herptiles; to Stuart Worth of Euro-Rep Ltd for donating much of the equipment featured in this handbook; to Simon King of King's Reptile World for loan of livestock and general advice; and to Virginia Cheeseman for her expertise on invertebrates.

**Dedication**

For Jack and Georgia

# EXOTIC PET HANDBOOK

A guide to buying, caring for and breeding unusual pets

**David Manning**

First published in hardback in 1998 by
HarperCollins*Publishers*, London

First published in paperback in 2003 by
Collins, an imprint of
HarperCollins*Publishers*
77-85 Fulham Palace Road
Hammersmith
London W6 8JB

The Collins website address is www.collins.co.uk

Collins is a registered trademark of
HarperCollins Publishers Limited

09 08 07 06 05 04
7 6 5 4 3



A catalogue record of this book is available
from the British Library.

ISBN 0 00 714258 7

**Please Note**
While every reasonable care was taken in the compilation of
this publication, the Publisher and Author cannot accept
liability for any loss, damage, injury or death resulting from
the keeping of exotic pets by user(s) of this publication,
or from the use of any materials, equipment, methods or
information recommended in this publication or from any
errors or omissions that may be found in the text of this
publication or that may occur at a future date, except as
expressly provided by law.

Colour reproduction by Colourscan, Singapore
Printed by Printing Express Ltd, Hong Kong

# CONTENTS

# Introduction

The interest in keeping reptiles, amphibians and invertebrates as pets is growing at a considerable rate every year. More information is becoming available about these animals' requirements in terms of successful management and captive breeding. They are also becoming more readily available from local pet stores. As a result of this knowledge and the availability of many unusual species, more and more people are being drawn to keeping these fascinating creatures.

A reptile, amphibian or invertebrate may not seem an appropriate choice of pet to many people. However, many of these creatures are beautiful to look at and relatively easy to keep. They can also be tamed and are extremely popular amongst younger people who wish to keep a more unusual type of pet. Many of the animals featured in this book are incredibly marked and beautiful in their own way, and are fast catching up with small mammals in terms of popularity and suitability as pets.

Many species are easy to maintain. However, there are obvious responsibilities that need to be considered before you purchase one of these creatures. The housing and needs of these animals need to be thought about, and for younger children the adult or guardian must ultimately be responsible for the welfare of the chosen species.

This book offers guidance and practical advice to parents and children on choosing a more unusual family pet. The positive and negative aspects of keeping these creatures is also discussed. On the whole, however, these animals are low maintenance, allergy-free, alternative pets. I have deliberately concentrated on those species that meet the following two criteria:

- They are widely available in the trade or specialist pet market, preferably from captive bred stock. This helps to ensure that these creatures are free of internal or external parasites; imported specimens are more difficult to maintain and are not very hygienic to keep as a family pet.

- They are suitable for the novice keeper. I have tried to promote the care of the most interesting and easy to keep species, which are not too demanding on time and space. At the same time, however, I have also included some of the most sought after species, such as chameleons and iguanas.

Each section lists species in an 'ease-of-care' order. The easiest to keep appear first, graduating to those that are more demanding in terms of equipment requirements, space and maintenance time.

## CONSERVATION NOTE

Reptiles, amphibians and invertebrate numbers are rapidly decreasing due to habitat loss, pollution, global warming, soil erosion, deforestation and natural disasters such as volcanic eruptions and floods. Many herpetologists, both in zoos and hobbyists at home, are discovering much needed information about these species. Captive breeding is one of the great hopes for the continual survival of many creatures and it is hoped that they may one day be reintroduced into the wild if habitats can be saved.

▲ **Exotic pets, such as these Leopard Geckos, are relatively easy to maintain.**

The collection of certain animals from the wild and their import and export between countries may be illegal. The Convention of International Trade in Endangered Species (CITES) is an international agreement that regulates and monitors trade of wild animals and plants, including herptiles. It is impossible to obtain many endangered or threatened species without a licence from a CITES signatory country. When purchasing a pet, make sure you buy from a reputable supplier.

Having said that, I hope that you enjoy your involvement with these fascinating creatures, and that you also try to protect any native habitats and those species that live within them. And finally, remember that pet keeping involves responsibilities as well as rewards.

# Acquiring your Exotic Pet

Before purchasing your pet, check with your local pet store, reptile club or Department of Wildlife to see if any restrictions exist on keeping a particular species, and whether a permit is required. Depending on which part of the world you live in, local and/or national legislation may affect your choice. You should be able to find information on relevant regulations at your local library. At the time of writing all the species featured may be kept in the UK without any legal restrictions.

## FIRST STEPS

Read all about the animals and their care and then decide on a species that will suit your circumstances (see Box on p.9). Next find a supplier. It is easy to go to a pet dealer to buy your pet and its home at the same time. However, you then have the problem of trying to set up a vivarium or aquarium quickly before your pet gets too cold, dry or hungry. The best thing to do is to **plan in advance.**

First consider which species you wish to keep and maybe place a deposit on the animal of your choice. Next, purchase the accommodation and equipment; take these home and create your setup. Do not rush things – a vivarium may take several hours to arrange correctly. Over the next few days monitor the temperature and humidity to ensure that you have created the right environment for your pet. With an aquarium or semi-aquatic vivarium it is essential to establish the tank for 1-2 weeks before introducing your pet. This allows the water to age/mature. If you are using a filter, beneficial bacteria will start to establish themselves and will help to improve water purity.

Now you may consider selecting your pet. To increase the chances of keeping your pet healthy and in top condition you will need to select a healthy individual to begin with, preferably one that is captive bred and that is also an established feeder.

*Look for specimens:*
- that are bright-eyed and alert-looking
- that are good feeders
- that do not bolt or flee

*Avoid specimens:*
- that have a dirty vent – caked-on faecal matter is a sure sign of problems
- that cannot support their weight or that walk in an uncomfortable manner
- with very little fat reserves around the base of the tail
- with bits of shed skin still stuck to the body
- that are too small or the runt of the litter

Ask your supplier lots of questions. If he is a good one he will expect you to ask questions and should be able to assist you in selecting a suitable individual.

It is very likely that many of the baby snakes, lizards or frogs you are going to purchase will have been reared in what, at first sight, may seem like very unsuitable conditions – plastic boxes, often clear, measuring only 25 x 10 x 15cm (9.75 x 4 x 5.75in) with tiny air holes drilled into the sides or lid for ventilation.

▶ **Small lizards are a popular first choice for the novice keeper.**

Each box, containing a baby snake for example, will be lined with paper and equipped with a tiny shelter and small water dish. Providing it is warmed with a heater pad or cable, this is the ideal rearing container in which to 'nurture' a young snake and to raise it as a tame pet. To suddenly put such a fragile animal into a large vivarium can cause great shock to the young snake. It may feel vulnerable and refuse to feed or it may hide itself away, never to be seen again. It is best to leave the young snake in its rearing container and when you feel that the time is right – once it has grown a bit and feeds regularly – transfer it into a larger, more permanent setup.

## WHICH TYPE OF ANIMAL IS RIGHT FOR ME?

Before choosing a particular species, ask yourself the following questions:

**Q:** Do I want a pet I can handle sometimes?
**A:** You may wish to consider a Blue-tongued Skink or a Leopard Gecko.

**Q:** Would I prefer a pet that is simply great to watch in its environment?
**A:** Think about keeping an Axolotl, a clawed frog or a community of anoles.

**Q:** Are exotics easy to feed and will I mind feeding a predator its prey?
**A:** You may prefer a millipede to a mice-eating snake.

**Q:** Do I have enough room to accommodate my pet once it becomes a fully-grown adult?
**A:** Iguanas require a lot of space and attention and become very large once adult. A Tarantula is more suitable if you are short of space because it will always live in a small vivarium.

**Q:** Can I afford the equipment and foods necessary to keep my pet happy and healthy?
**A:** Chameleons require heaters, thermostats and fluorescent tubes. A Stick Insect, however, is much cheaper to buy and easier to maintain.

**Q:** Will I mind getting bitten occasionally?
**A:** Most snake owners are bitten eventually. The bite will not hurt a lot, and for most species, it will require little more than a plaster. If you handle exotics, there is always a possibility of being bitten. Snakes and larger lizards are the most likely to bite when mishandled, while still untamed or if they mistake your finger movements for prey. Tree frogs, salamanders, newts and most stick insects are virtually unable to ever hurt you in this way. A Tetanus booster or inoculation may be required if your skin is punctured.

**Q:** Am I responsible enough to care for an animal for the whole of its life and do I have the time to administer to its needs?
**A:** Any form of pet keeping is a responsibility. You must be responsible for an animal's care and maintenance. You will hopefully enjoy its company and get much satisfaction from nurturing a living creature, admiring its beauty or watching its behaviour. It is fun to handle your pet and to watch its general behaviour. It is probably less enjoyable cleaning out the faeces or obtaining and keeping a selection of live foods. If this does not sound appealing, then maybe you should opt for a low-maintenance animal, or reconsider your decision to acquire a pet.

## BENEFITS OF KEEPING AN EXOTIC PET

**1** You are unlikely to be allergic to an animal that has no fur or feathers.

**2** Reptiles, amphibians and invertebrates may be kept in locations that prohibit the care of other species, for example, a flat with no garden.

**3** They are generally silent and do not require regular walks, annual veterinary inoculations or grooming.

**4** Most species are easily maintained, requiring only a 10 minute feed each day and a cleaning regime that only takes 10 minutes per day, or an hour or so a week.

**5** Most species are happy to live within their unit and are not hemmed in like some animals. Species such as Milk Snakes prefer small, secure accommodation – larger units may cause them stress.

**6** Although a relationship may be established, your pet snake, unlike a dog, is not going to become emotionally dependent on you and it will not be distressed if you move house, or it has to be passed to another owner.

**7** These creatures will give you an interest in the world of natural history, conservation and a respect for all living things.

**8** Unlike a dog, your pet herptile or invertebrate will not feel lonely if you go to work or school during the day.

## WHERE TO OBTAIN STOCK

Joining an herpetological society is an excellent way to obtain stock and to get up-to-date information and opinions on suitable species. Ask friends, surf the net and visit your specialist pet dealer. Your local phone book and pet magazines are also useful sources of information.

## SCIENTIFIC NAMES

Confusion may arise when common names are used for particular species. For example, in the UK the small African Python is called the Royal Python, while in America it is known as a Ball Python. To avoid confusion, zoologists use Latin names to describe animals and plants. A code was developed in the 1700s by Karl von Linné. In this system, the Royal Python is called *Python regius* and the Latin name is written in italics and usually follows the common name.

When purchasing a pet it is best to ask for it by the Latin name. You can then find further information in books without getting confused about the species you wish to know about.

## HANDLING

Many of the featured species are amongst the most handleable of all herptiles. However, some species are enjoyed visually and once in the vivarium they are only handled when removed for cleaning.

When you handle your pet, close all doors and windows so that you do not lose it if it takes flight. Keep other pets and potential predators, such as cats, out of the vicinity. Minimize the danger of a fall by sitting on the floor. Remember that your pet is not a toy; children should handle pets only under adult supervision. Juvenile lizards or snakes should only be handled occasionally and for short periods until they become tame or conditioned enough to sit still and not to rush off. Feeding your pet after a handling session is a good way to condition it to accepting human contact as a stimulus.

Do not handle animals after they have eaten or when they are about to moult or slough. If transporting your pet, it must be comfortable and kept at a temperature and humidity level that will cause it no suffering – use an insulated container.

### AMPHIBIANS

The mucus-covered skin of many amphibians protects them from bacterial infections and makes them difficult to handle. It is best to handle them with wet hands. Some amphibians have toxins in their skin – if you have any open wounds, handling them may allow toxins to enter your body. Most amphibians can be held in cupped hands or scooped up into small containers for transportation. These can be filled with damp sponge to provide moisture. Transport totally aquatic species in water.

### LIZARDS

The dry skin and docility of the featured lizards make them ideal candidates for more intimate contact – they are amongst the most handleable species in this book. However, they should not be lifted by the tail – their ability to 'drop' them is well known.

**▼ Careful handling will allow you to enjoy close contact with your pet. Amphibians like this tree-frog need to be gently handled in a moist hand.**

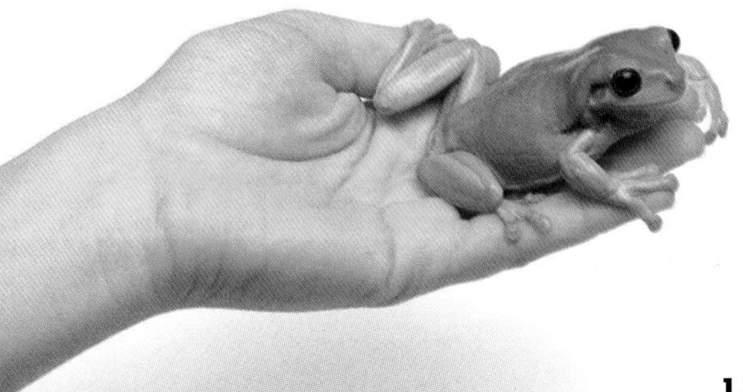

### SNAKES

Small snakes are easily supported in the hand; most will wrap around your fingers or wrist to feel comfortable. Never dangle them from the neck or tail, but support them along the length of the body. Heavier species should be held firmly and their weight supported. Untamed snakes should not be handled by the novice keeper, as they need proper restraining. Transport your snakes in cloth bags or tied pillowcases. Smaller species or juveniles may be transported in their plastic rearing boxes.

### INVERTEBRATES

Many invertebrates are delicate and need careful handling. Do not disturb them if they are moulting. Caution is advised when handling at all times because most species have tiny hook-like claws that easily get tangled up with clothing or may snag a small child's skin. If in doubt, do not handle.

## KEEPING A PET DIARY

All sorts of information can be recorded in a pet diary. Start with details of its size/weight when purchased, then record feeding habits, humidity, temperature, shed skins, activity, etc. Here are some of the things you may wish to record:

- Common and scientific names
- Date of birth (or purchase) and origin of pet
- Length and/or weight
- Food eaten - 'R' for Refused, 'A' for Accepted
- Medicines or vitamin supplements given
- Toilet habits
- Breeding details – mating, births, egg laying, etc.
- Shedding skin/moults – when and how often

If your pet becomes ill, a diary may help you or your vet to treat it better since you have a record of its life history. It will also help anyone who may have to look after it for you.

# Accommodation and Equipment

To create the right environment for your pet you must have an understanding of the types of accommodation and equipment that are available. The correct choice of these items should ensure that your pet enjoys a long and healthy life.

## HOUSING

A vivarium or aquarium can be a beautiful and enchanting miniature natural habitat which both you and your pet can enjoy. Keeping an exotic pet can also offer an insight into the natural world which, without our interest and involvement, is easily misunderstood or forgotten.

Depending on the eventual size your pet reaches as an adult, you may be looking to increase unit size gradually, or you may keep one unit for your pet's entire life. I generally encourage larger vivaria for most pets – it allows you to keep more individuals together and even to establish a community setup, giving you more scope to create an attractive environment for you and your creatures to enjoy.

There is a range of suitable containers available and these are divided into three main categories.

**Glass** Glass is the most versatile option for housing. You can keep any animal mentioned in this book in a glass vivarium or aquarium. Many come equipped with sliding doors and a lock can be fitted easily for added security. For extra protection the doors and larger surfaces can be made of toughened glass. Some vivaria also have mesh ventilation panels allowing free movement of air. Glass is easy to clean and provides good all round visibility. It is important to provide adequate cover and hiding places so that your pet does not feel exposed. These vivaria can be custom made to suit any animal's particular requirements.

**Wood** Some wooden or chipboard vivaria and cabinets can be purchased as flat packs for easy assembly at home. One drawback is that if you are keeping a species that requires high humidity, the moisture will eventually seep into the wood and damage the unit. Although popular, and being widely stocked in pet centres, they are, in my opinion, the least aesthetically pleasing of all the three options.

**▶ This glass vivarium contains many of the features required for successful herptile maintenance. The spotlight creates a hot spot at one end and a range of climbing places and retreats, seen on the left, help to create a temperature gradient within the vivarium.**

▲ **A selection of plastic pet containers in a variety of sizes. Whatever accommodation you buy, make sure it is the best quality you can afford.**

**Plastic** Several large pet product manufacturers make plastic pet containers. They are useful for rearing juveniles or for storing and breeding insects, invertebrates and live foods. They are colourful and cheap, but few of them are secure, large or sturdy enough for long-term maintenance of snakes or larger lizards. Unless you use only a heater pad, most of the plastic units have little or no potential for attaching a spot light, thermostat or full spectrum tube. They are especially suitable if you are keeping your pets in a room that is already heated. The acrylic material deteriorates with cleaning and is never as clear as glass. However, they are easier to move about than wood or glass and are intrinsically safer than glass. Most pet stores stock them.

Whatever option you choose, the golden rule is to buy the best quality accommodation and equipment you can afford. Do not try to save money on what may be the most important aspect of your pet's welfare. In common with many other pets, its accommodation and equipment will often cost more than your pet. Good equipment will last a long time and will provide a suitable environment beneficial to the long term well-being of your pet.

## LIGHTING AND HEATING

In the wild, most diurnal herptiles are exposed to sunlight and the healthy growth and development of many species is directly influenced by it. Heat is also essential for many of the featured species, since they naturally inhabit warmer climates.

### LIGHTING REQUIREMENTS

In the vivarium, you must try to offer the correct conditions for your pet to thrive in, and this means providing suitable artificial light.

Light is also crucial for many aspects of behaviour in reptiles and amphibians. Some species, for example, like to 'bathe' or bask, not just in the heat but also in the light. The amount of light influences the perception of seasons – more in summer and less in winter. The number of daylight hours and the length of a season can be a very important factor if you want to breed your creatures.

Nocturnal animals are most active in low light conditions. Crepuscular species, like many toads and geckos, are most active at dawn and dusk, when the light is reduced. Other species, such as the Bearded Dragon and Veiled Chameleon, will sunbathe in brilliant daylight at temperatures up to 38°C (100°F), but they still need a night-time dark period for sleep.

Herpetologists have known for many years that Vitamin D3 is a crucial requirement for the absorption of calcium, a vital ingredient for healthy bone development in many reptiles and amphibians. Without exposure to the sun's ultraviolet rays, the animals will suffer and Metabolic Bone Disease may develop – probably the most frequent problem seen by vets amongst this group of animals.

Fluorescent tubes provide light with a minimal amount of heat. Incandescent light bulbs and spotlights provide a high level of heat, depending on wattage, but provide none of the beneficial ultraviolet wavelengths. For temperate species, such as European Treefrogs, a fluorescent tube is ideal for providing light with only the minimum of heat to prevent your pets from overheating. For many species, especially tropical ones, both fluorescent tubes and incandescent light bulbs/spotlights are necessary. The spotlight, or incandescent bulbs, should be thermostatically controlled to ensure that the correct temperature is maintained.

Do not place your vivarium near a window to provide extra sunshine for your pet. The useful ultraviolet rays will not penetrate glass or other barriers and may well turn your vivarium into an oven with direct sunlight lifting the temperature to a fatally high level.

In essence, a full spectrum light is at the very least beneficial and often essential for herptiles that need to bask and for many diurnal species. Fortunately, with the herpetological trade constantly growing, a good selection of light tubes are now available to suit a variety of species which require UVA and UVB light.

Spotlights are very useful for general lighting and provide hot basking spots, essential for thermoregulation. When 'night' falls after 12-16 hours of 'daytime', the temperature of a vivarium may drop quite significantly without the additional aid of a heater pad (depending on the species). Nocturnal species can be seen better by the human observer if a blue or red bulb is used, in conjunction with a daytime spotlight, on a second timer for night. Often the background light in the room is sufficient to enjoy your pet, unless you have particularly large or well-planted setups.

## HERPTILES AND THERMOREGULATION

A temperature gradient is essential for the successful maintenance of most herptiles. They thermoregulate, or move between different temperature zones, to warm up or cool down as required. Without adequate warmth they cannot feed, digest food, move about or produce sperm. At different times of the day or night they will seek out an environment that suits their needs. You must provide these different environmental conditions within the vivarium. A good temperature gradient can be established using a thermostat to control temperature.The diagram below shows the effect of temperature on most herptiles. Many prefer a vivarium with temperature zones within the generally preferred range.

| | CRITICALLY COLD | | | GENERALLY PREFERRED RANGE | | CRITICALLY HOT | |
|---|---|---|---|---|---|---|---|
| °C | 0 | 4 | 10 | 20 | 35 | 45 | 50 |
| °F | 32 | 37 | 50 | 68 | 95 | 113 | 122 |

▲ **Keep animals at their preferred temperature. If it gets too hot or cold they may suffer or even perish.**

## EQUIPMENT

Your choice of pet will dictate what equipment you require to create the right environment. Every animal in this book may be housed successfully using a selection of the equipment described.

**Thermometer** One or more should be placed strategically near the main heat source and midway towards the cooler end of your vivarium to make accurate temperature checks. A temperature gradient allows the herptile to thermoregulate effectively. Air temperature needs to be measured for arboreal species and ground temperature for terrestrial species.

**Hygrometer** This instrument measures humidity, a very useful addition to any environment that needs regular monitoring.

**Heater pads** An excellent range of low, medium and high wattage pads are available. Low wattage heater pads are ideal for heating invertebrates, or you may use higher wattage pads in conjunction with a thermostat. Never cover more than half the ground area in the vivarium with a heater pad. The pads need regular checking, since their effectiveness can deteriorate over time.

**Ceramic heaters or infra-red bulbs** These are powerful heaters that are excellent for creating hotter basking areas for Bearded Dragons and Veiled Chameleons. They do not provide any light. They must be thermostatically controlled and animals and keepers must be protected from contact. Read all instructions carefully.

**Thermostats** Essential equipment for the regulation of temperature. Use them in conjunction with heater pads or bulbs. Different thermostats are available to control pads and bulbs.

**Fluorescent tubes** A range of products are available to suit the needs of most species. These tubes can provide essential UVA and UVB light and will also help to promote good plant health. Sizes range from 45-150cm (18-60in) and each tube requires a starter pack, a device that regulates the power needed by the tube.

**Light bulbs** These are useful for heating and lighting purposes, but have no ultraviolet light content. Vivarium animals and furnishings should be protected from contact with them. For proper heat control they are best regulated with an appropriate thermostat.

**Spotlights** They are most useful for providing direct heat and light to a basking area and will shield much of the glare of normal bulbs from the viewer.

**Water bowls** Try to keep them level with or lower than the surrounding ground level. You can use almost any suitably sized dish or jam jar lid. The manufactured types look more natural and have rough edges to give reptiles a better grip for getting in and out of the water. Keep the bowls clean and replenished with fresh water. Remember, some species prefer large bowls for bathing and others only drink from water droplets.

**Hide rocks and wood** Invaluable for giving many herptiles and invertebrates a sense of security – a rock to crawl under. It is better to have too many than too few. Terrestrial animals want ground hides, while arboreal species prefer higher hides attached to the side of the vivarium. Manufactured hides are easy to clean. Cork bark is attractive and can be purchased in logs that are easy to cut or break into smaller pieces. Avoid stacking heavy rocks – they may collapse and injure a pet that is digging.

**External power filter** For optimum reptile and amphibian health, water purity is essential. The larger external power filters will heat the water as well as keeping it clean and flowing. I cannot overstate how important water quality is, especially in larger planted vivaria for semi- or totally aquatic creatures. Ideal for any herptile requiring higher humidity or semi-aquatic conditions, is water directed over a rock or log strategically placed in reach of thirsty animals. Using de-mineralized water will reduce the amount of scale gradually deposited on glass.

**Naturalistic backgrounds** These long pictures of natural landscapes are effective for any setup and come in a variety of options, from tropical rainforest to mountain views. They are cut to the required length and may be simply affixed to the outside of a glass vivarium or inside a wooden one. If used inside, take care attaching it as sticking tape can tear a reptile's skin.

**Plastic and silk plants** Available in a range of colours and leaf forms, plants are very useful for providing cover, resting and climbing areas and for making a setup look more attractive. They can be washed and disinfected easily. A small pile of plastic plants sprayed with water will create a humid micro-climate that will be appreciated, even by desert species – especially when sloughing, and will help to stop dehydration. Ensure plastic plants and other furnishings are never sited too near to spotlights, light bulbs and other heaters.

**Real plants** Real plants are good for larger vivaria, especially those with full spectrum lighting that will help plants to thrive. Sturdy plants, such as Philodendron, are ideal for chameleons, whereas a spider plant would suit anoles or other light-weight species. There is no need to plant out a nicely arranged setup for larger, more boisterous species, such as the Blue-tongued Skink or Iguana, since they will soon destroy or flatten your efforts.

**Books** Books are a great source of information – visit your library or local book shop. Many animal societies produce specialist publications and scientific papers giving more in-depth or specific information on a particular species.

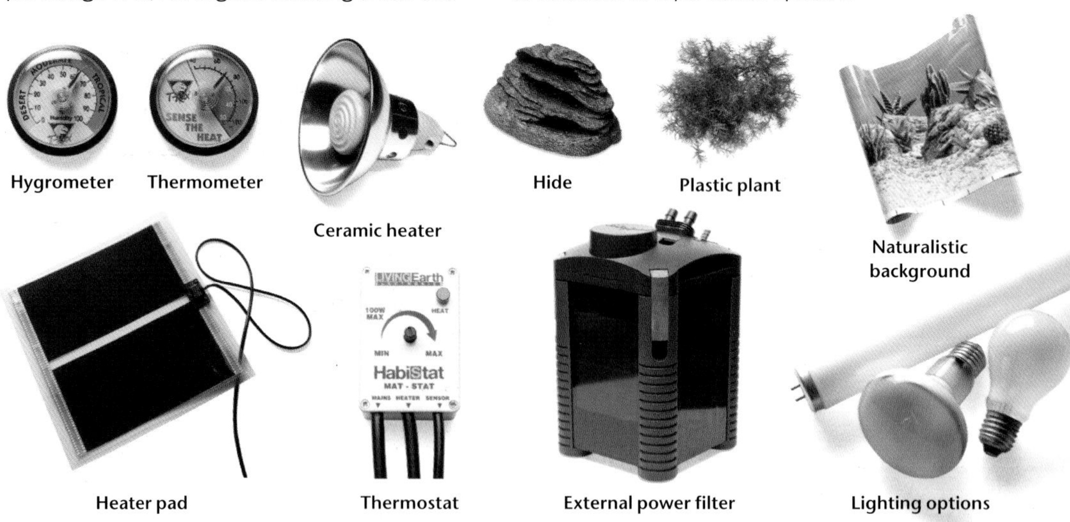

Hygrometer Thermometer Ceramic heater Hide Plastic plant Naturalistic background Heater pad Thermostat External power filter Lighting options

It goes without saying that you must read the instructions carefully on all equipment and follow normal electrical safety rules. If in doubt, contact a qualified electrician. All electrical equipment should be checked regularly.

All the above information and advice reflects the current state of knowledge in the care and accommodation of amphibians, reptiles and invertebrates. New developments are constantly occurring which help us to understand better the needs of herptiles and new products to assist the keeper are constantly appearing on the market. For the very latest information, you should consult your specialist society and specialist magazines.

## SUBSTRATES

A large variety of substrates are available and your choice will depend on the needs of the pet and your aesthetic preference. For ease of maintenance, many breeders and herpetologists keep lizards and snakes in very bare, laboratory-style setups, with little more than newspaper as a base. This may not affect the happiness of the animal at all, but I prefer more naturalistic displays incorporating a mixture of substrates and furnishings.

Below are listed some of the more popular options, but as long as the substrate is clean and free from the introduction of pests, parasites or toxins, many natural items found in your locality – twigs, leaves, stones and gravel – can be utilized.

**Play sand and gravel** Useful for desert species, a mixture of sand, gravel and wood chips will enhance the design of most vivaria. Some specialist reptile stores sell coloured sands – highly decorative for desert-style setups.

**Wood or bark chips** There are a wide variety available. Avoid the finest types because the particles may be taken in with foods, causing digestive problems. This is a very common problem with captive lizards, especially juveniles.

**Leaf litter** Attractive seasonal leaves, conkers, pine cones, etc., can all enhance the appearance of a vivarium. These are best frozen overnight to kill any resident pests.

**Reptile grass** This is a green matting that is easily washed and is very versatile. It can be placed on the ground, or on the side of the vivarium for arboreal species.

**Paper** It is easy to clean and replace, but not very pleasing to the eye. Ensure your vivarium pet is not under-stimulated in a bare or barren vivarium.

**Vermiculite** This is a natural product that is excellent when wetted (and excess moisture squeezed out) to provide an incubating medium for most eggs. It can be used as a substrate for spiders and other insects.

**Purlite** These are compressed, clay-like balls used in horticulture. Ideal for semi-aquatic species. It retains moisture well but is considerably heavier than vermiculite.

# Foods and Feeding

All living creatures require foods to grow and thrive. Some are specialists, and eat only one type of food. Herbivores, for example, are animals that eat only plants and fruits. Some animals are less specific and will eat a variety of foodstuffs – other animals, insects, plants and fruits – and are known as omnivores. Animals that feed only on insects are known as insectivores and those that only feed on other animals are classed as carnivores.

You will need to determine which group your pet belongs to and then offer it the right selection of foodstuffs, in sufficient quantity and quality, to satisfy its needs. Whatever the diet, the food should supply that species' requirements in terms of fats, proteins, carbohydrates, vitamins and minerals. You also need to take into account that feeding can be an important behavioural stimulus. Many insect-eating herptiles, for example, spend much of their life hunting for and catching food. Getting the feeding and the foods right for each animal is all part of successful pet management.

Fresh water should be available in shallow, easily accessible bowls for most terrestrial species. Mist spraying is beneficial to many herptiles and invertebrates, since they often prefer to drink from droplets rather than bowls.

## HERBIVORES

Many birds, mammals and invertebrates are herbivores, as are most tortoises and some lizards. Many species, however, do supplement their diets with protein when available. Some is unintentionally consumed with plant matter, some intentionally sought.

There is a wide variety of fruit and vegetables that can be shop bought, or grown in a pesticide-free garden or allotment, that will suit herbivores. Amongst the most suitable are lettuce, broccoli, tomato, pear, apple, seasonal berries such as raspberries and blackberries, dandelion flowers and their leaves and roots, spinach, carrot, Brussels sprouts, cooked potato, cabbage, grass and non-poisonous tree leaves, flowers and seeds.

A variety of complete foods are also available and they are balanced meals that have added vitamins and minerals. Personally, I believe that you can do no better than provide your pet with a range of fresh rather than processed foodstuffs. For convenience, however, they are handy and not subject to seasonal availability like fresh produce. These ready meals are particularly useful if a friend or relative is caring for your pet in your absence.

## OMNIVORES

Creatures that feed on plants and other animal species are considered to be omnivores – this category includes humans. Omnivores covered in this book, such as the Bearded Dragon, Blue-tongued Skink and Veiled Chameleon, will be satisfied with a combination of the foods listed under the herbivore, insectivore and carnivore sections.

Individual preferences vary but Blue-tongued Skinks, for example, are generally partial to snails offered in their shells; an adult specimen will probably enjoy two or three snails and a quarter of a banana at one sitting. Insects are full of protein and few lizards will pass up the opportunity to eat these nutritious titbits, even those species considered herbivorous.

## INSECTIVORES

Due to interest in keeping and breeding herptiles, the range, quality and availability of live foods has increased greatly over the last few years. If you cannot obtain foodstuffs from your local pet store, you may easily purchase them by mail order. Insectivores, and most omnivores, will enjoy a selection from the following live foods, listed in order of size.

**Locusts** (*Locusta migratoria, Schistocerca gregaria*) The fat-bodied, winged adults are ideal for large insectivores, although the spiky legs are not appreciated by all. The young are referred to as hoppers. Usually measuring between 1-4cm (0.5-1.5in), they are ideal for smaller lizards and frogs. Locusts tend to climb upwards, so are ideal for arboreal species.

**Crickets** (*Gryllus sp*) Probably the most widely available and inexpensive of all insect foods, they range from tiny pinhead-sized hatchlings to adults measuring over 3cm (1.25in) in length. Various species are now reared but I prefer the black, silent crickets. They are plumper, quieter and less likely than the brown-coloured house cricket (*Acheta domestica*) to take up residence in the house if they escape. The hatchlings are excellent for the tiniest frog or newly hatched lizard.

**Mealworms** (*Tenebrio sp and Zophobas sp*) Many species find these beetle larvae difficult to digest. Never use them exclusively, but only as part of a varied diet. Once they metamorphose into beetles, most reptiles will refuse to eat them.

**Waxworms** (*Galleria sp*) They are the larvae of a small moth and are also edible in their adult form. Waxworms are one of the most nutritious and easily digested of live foods; they are very useful for building up a weak, ill or stressed pet. Excellent to use when trying to tame your lizards and frogs, but don't use them too often as they are very rich. They are fast movers, which will wriggle into tiny nooks and crannies within the vivarium quickly – feeding by hand will prohibit this.

**Fruit Flies** (*Drosophila sp*) These little flies suit many baby lizards, frogs and invertebrates, such as praying mantids. They can be raised on fruit or special culture mixes. Since Fruit Flies quickly infest a location, you may prefer to order large flightless Fruit Flies, *Trinidadian sp.*

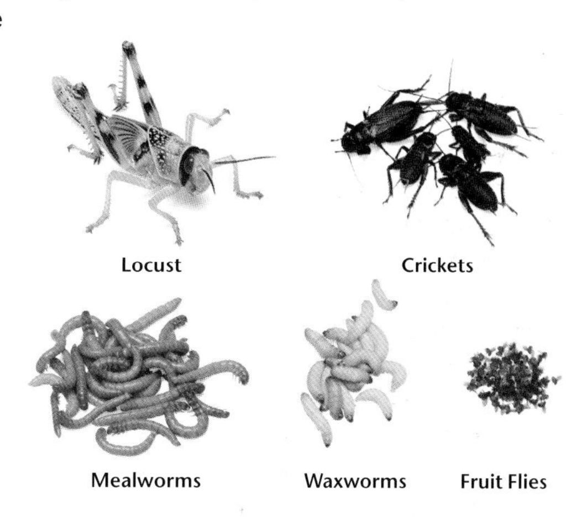

Locust — Crickets — Mealworms — Waxworms — Fruit Flies

## OTHER FOODS

The following additional foods are also suitable for both insectivores and omnivores.

### COLLECTING YOUR OWN

Commonly farmed live foods should be raised in hygienic conditions. One of the problems of collecting your own is that you always run the risk of introducing some parasite or unwanted guest. Having said this, one of the benefits of using wild caught live foods is that they are packed with naturally acquired vitamins and minerals.

**Snails** These calcium-rich creatures are a preferred food of Blue-tongued Skinks, many toads and small lizards.

**Worms** Only use earthworms purchased at a reptile shop or bred at home from purchased varieties. Some species can be toxic to amphibians and reptiles, especially those inhabiting compost heaps or rotting vegetation.

**Spiders** All spiders can bite, but most are not dangerous to humans; adult supervision is advised. Check the species before collecting.

Snails

Worms

Make sure that the creatures you collect are not dangerous to humans. Choose wild-caught prey with caution and avoid collecting in areas where pesticides or fertilisers are used. Most gardeners will happily allow you to collect snails and other 'pests' from their allotment or garden.

### AQUATIC FOODS

Foods for aquatic amphibians and rearing foods for animals at the larval stages of development can be purchased from good pet stores.

**Tubiflex** Red aquatic worms ideal for newts, Axolotls and for many animals at the larval/tadpole stages.

**Daphnia** Small crustaceans found in ponds.

**Brine shrimp** (*Artemia sp*) Purchased alive or as eggs, they require a salty solution that should be filtered before they are offered as food.

**Freshwater shrimp** Ideal for Axolotls, newts and Oriental Fire-bellied Toads.

**Pond pellets and flaked fish foods** Some Axolotls and African Clawed Frogs will readily accept pellets. Most tadpoles will munch on flaked fish food, a complete and nutritious foodstuff.

## CARNIVORES

Carnivore is a term largely used to describe flesh-eating animals. Within this book, the only true carnivores are snakes. For most of their lives snakes in captivity will accept dead, thawed out, rodents, and are known as 'defrost feeders'. The rodents may be purchased ready frozen. For humane reasons, your pets should not be fed live vertebrate food. Snakes consume their food whole, so there is no mess or leftovers and it would seem that their vitamin and mineral requirements are met totally by their small rodent diet. Supplements are therefore not generally needed, although you may add vitamin drops to the snake's water occasionally.

**Mice** (*Mus musculus*) The best and most popular foodstuff for snakes. Juvenile, hairless mice, called pinkies, are an ideal size for hatchlings and juvenile snakes. Furries, juvenile haired mice, are the next size up. Purchase them ready-frozen from your pet store or specialist herpetological supplier.

Adult Mouse　　Pinkies

**Rats** (*Rattus*) Small and medium-sized rats, preferably brown or black in colour, are a good sized food for Royals and larger Kingsnakes.

**Other rodents** Many smaller rodents will be accepted by a variety of snakes – gerbils are particularly favoured by Royals. All frozen rodents should be fed to your pet only once they are thawed and warmed. Should you breed your own, rodents must be kept and then killed in a humane way.

Some processed snake foods are now available. They are more expensive than rodents and I personally do not think they are necessary. Those who do not wish to feed animals to each other may prefer to buy foods that no longer resemble an animal or which are processed. I would suggest that if you do not like feeding a predator, you should consider a different pet.

Many species prefer to feed at dawn or dusk, or at least in low light conditions. Offer your pet a thawed-out rodent held by its tail, using long tweezers. It should be accepted readily if the snake is hungry and not nervous. For many snakes, and especially juveniles, it is worth leaving the food in a hide box overnight, where the snake can find it at its leisure and then consume it in peace. Uneaten food should be removed and disposed of – do not re-freeze it.

## SUPPLEMENTS

A range of products are now available for herptiles. The most important ones for many species contain added vitamins and minerals. Some new products even claim to provide the UVA and UVB requirements of basking species in a liquid form. The best advice I can give, as with all things, is to use all these products in moderation and at the correct dose.

## FEEDING TIPS

Some people prefer to feed their pet in a separate container to the one in which it lives. Plastic pet homes are ideal and can also be used to keep your live foods. With this method it may be easier to monitor how much is eaten by each pet. It may also train your pet to accept being handled first before receiving its food reward in its feeding box. It is important to keep and breed your live foods in hygienic and humane conditions. Care instructions should be available from your supplier or pet centre.

## WHEN ANIMALS REFUSE TO FEED

Species can stop feeding for many reasons. If you have had your pet for a while, it is likely that it is well fed, with built up fat reserves, and is 'full'. After a short break – a few days for a lizard to several months for a snake – it will continue to feed normally, so do not worry if your pet refuses food occasionally. However, there are times when you should be concerned about a lack of appetite:

**a)** If you have just acquired a pet and it has not eaten at all within the first week or so.
**b)** Your pet is getting thin or weak, or becoming less and less active.

If your pet is not eating, check the following:
**a)** Have you created the right environment for it? Is the unit warm enough? Is there enough water?
**b)** Have you offered foods of the right type and size? Is the live food hassling your pet? Is the food offered in the right place and at the right time of day?
**c)** Is your pet sloughing? Most species will refuse foods at this time but are hungry again afterwards.

If in doubt about your pet's health, consult a vet. Any records you have kept will assist you or any professional to determine if there is a problem.

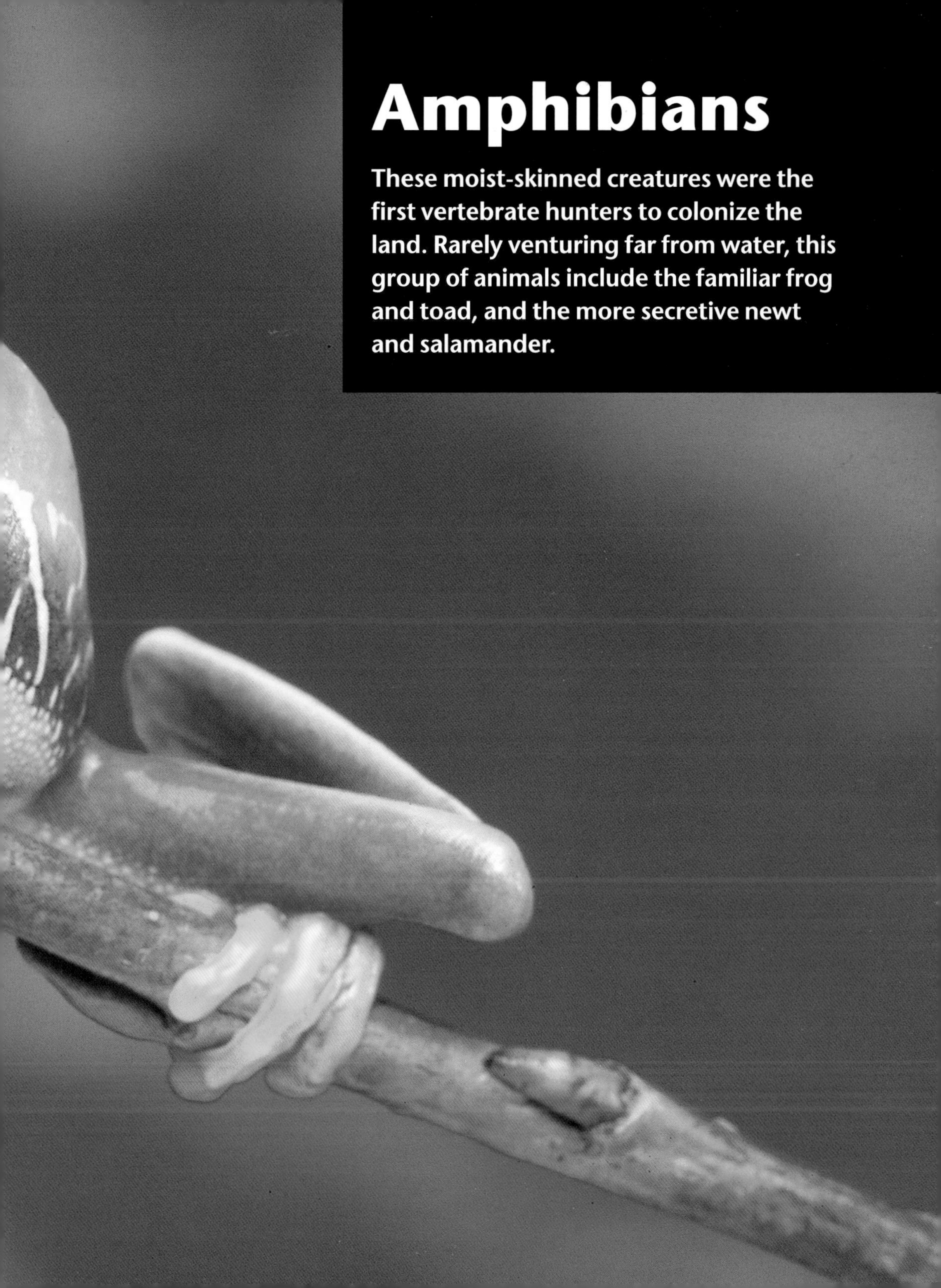

# Amphibians

**These moist-skinned creatures were the first vertebrate hunters to colonize the land. Rarely venturing far from water, this group of animals include the familiar frog and toad, and the more secretive newt and salamander.**

Amphibians ruled the earth before any reptiles or mammals even existed. They were the first predatory vertebrates to adapt to life on land. The word amphibian is derived from the Greek words 'amphi', meaning both, and 'bios, meaning life. As their name suggests, amphibians are suited to both life on land and in the water. However, few of them venture far from water.

Amphibians are ectotherms and they are entirely reliant on environmental warmth to maintain bodily functions. Most species from temperate climates hibernate in winter to avoid lethally cold temperatures. Species from tropical climates aestivate to avoid lethally hot conditions.

Amphibians have delicate, moist skin which is covered with a protective mucus coating. This water-permeable skin offers very little protection from external toxins and chemicals, making amphibians very sensitive to their environment. They slough regularly and the nutritious, sloughed skin is generally eaten. Many amphibians display wonderful colours, either for the purpose of camouflage or as a warning to potential predators to show that they possess toxic skin.

All adult amphibians are predators. They feed mainly on spiders and insects and the larger species are even capable of eating small reptiles and mammals. Foods are swallowed rather than chewed. To aid in this process frogs, for example, possess tiny peg-like teeth to hold the prey before swallowing. As hunters, the eyes of amphibians are essential tools and need to be kept clean and wet. The eyes are covered with a fine membrane and blinking causes this membrane to wipe clean and moisten the eyeball in one action.

Amphibians generally lay soft, jelly-covered eggs into an aquatic environment. Thousands of eggs may be laid, but mortality is very high and fewer than one percent of these are likely to survive to breeding age in the wild. The eggs hatch into

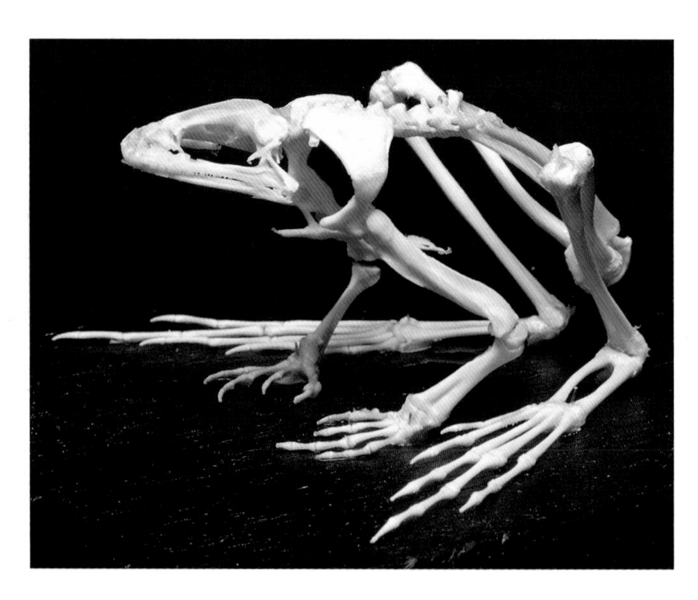

**◀ The skeleton encases several organs found in all vertebrates – the liver, lungs, heart and stomach. In this skeleton of the Common Frog, *Rana temporaria*, the large leg bones are clearly visible which, powered by strong muscles, enable frogs to leap great distances. Frogs also possess tiny, peg-like teeth to hold prey before it is swallowed.**

larvae, also known as tadpoles, and these eventually undergo a shape changing metamorphosis into the adult form. These young adults then leave the water and adopt a more terrestrial lifestyle.

There are some species of amphibian, however, who do not follow these breeding rules. The unusual Axolotl is the classic exception. This creature has developed the capacity to breed whilst still in the larval, or tadpole, stage. This ability is called neotony.

Amphibians are in decline globally – pollution and habitat loss are the major factors. Since their skin is water-permeable, it offers them very little protection from man-made chemicals and the loss of breeding ponds results in populations being decimated. However, due to their high breeding capacity, it is possible to effectively repopulate areas with captive-bred stock over a relatively short period of time, providing the environmental conditions have improved.

**▲ Smooth, moist skin is a characteristic of amphibians, such as this Fire Salamander. Unlike frogs, newts and salamanders are silent, unable to croak or call. They are also deaf, since they possess no ears. They are nocturnal creatures and actively hunt worms, slugs and other invertebrate prey.**

## AMPHIBIAN CLASSIFICATION

| Order | Description | No. of Species |
|---|---|---|
| ANURA | Frogs and toads. The best known of all the amphibians. Possess legs, but lose their tails once adult. This is the largest order, and contains some of the most widely-kept amphibians. | 2,600 |
| URODELA | Newts and salamanders. Limbs and tails are both present in their adult forms. They do not have ears and sight is the major sense organ for finding food. | 300 |
| APODA | Caecilians - long worm-like amphibians. These are not covered in this book. | 150 |

# AXOLOTL

*Ambystoma mexicanum*

**Found in the wild only in the canal system of the former Lake Xochimilcho in Mexico, these strange amphibians are easy to keep and make attractive, if unusual, aquatic pets.**

Although officially an endangered species, Axolots have been available from captive-bred stock since the 1830s. They are available in four colour varieties. Black, or dark grey, is the most common colouring but albino, harlequin, and, more recently, a golden form, have all been captive bred and are available to the pet keeper. Growing to 30cm (12in), an Axolotl should live for 10-12 years in captivity. However, older specimens of some 20 years have been reported.

An unusual feature of Axolotls is that they do not go through a normal amphibian life cycle in which eggs laid in water by the adult hatch into tadpole-like 'larvae', and then metamorphose into adults. At this stage the animal frequently leaves the water to take up a more terrestrial lifestyle. Axolotls have foregone the need to undergo metamorphosis. They remain in the aquatic, larval stage and reproduce without the need to become adult. The ability to do this is called neotony.

An Axolotl can only change into adult form when the hormone thyroxine is introduced into the diet or when there is an increase in the iodine levels in the water. The Axolotl gradually loses its gills, the tail reduces and, upon leaving the water, becomes a Mexican Salamander. Metamorphosis will not normally take place without this special treatment.

## CREATING THE RIGHT ENVIRONMENT

A perfect home for a single Axolotl is an aquarium measuring 60 x 38 x 30cm (24 x 15 x 12in). The water should be aged tap water, i.e. left standing for 48 hours. Make sure the water is roughly neutral (pH 6.5-7.5) and maintained at a temperature of between 10-25°C (50-77°F). A heater is therefore not usually required. Water depth can be at least as deep as the length of your Axolotl, but specimens missing a limb or two may prefer shallower water until these have fully regenerated.

**▶ A healthy Axolotl will ideally have four limbs, a complete tail and should hold itself stable in the water. Three branch-like gills on each side of the head enable the Axolotl to breathe underwater. The cooler the water, the smaller the gills, since more oxygen is present at lower temperatures.**

### AMAZING FACT

The Axolotl is much studied due to its amazing powers of regeneration. A lost or damaged limb will regrow over a period of about 8 weeks. The new limb will be just as good as the old one and full flexibility will be retained. Occasionally, due to the extent of the original injury, the regenerated limb may regrow at an unusual angle.

◀ **This aquarium provides spacious accommodation for a pair of Axolotls. It is decorated with pebbles and rocks, and the water is cleansed using an internal filter. The water level is kept deliberately low; it is about as deep as the Axolotl is long. Plastic plants are an attractive and robust addition.**

Axolotls will soon ruin a delicately planted aquarium, so they are best kept in a relatively plain setup. Gravel, pebbles and even fine sand are suitable substrates. Adding larger rocks will provide visual appeal to the aquarium and will give your Axolotl some platforms upon which to rest. An external or internal filter is recommended to clean and aerate the water but it is not essential. Weekly cleaning of all the furnishings and partial changes of the water is an acceptable, although more time-consuming, alternative. Lighting the aquarium either from the side or from underneath makes a very attractive display.

Generally speaking, Axolotls are best housed separately, since they have a tendency to eat each others' limbs. These do regrow eventually, if not as perfectly as their originals. Keeping these creatures separately does not usually cause problems, since they seem totally unconcerned about the lack of company.

**HANDLING**

Axolotls are aquatic animals and do not like frequent handling. When necessary, they can be moved by scooping them into a net or by gently cradling them in the hands. Always transfer them between waters of similar age and temperature so as not to cause them to go into shock.

▲ **When your need to clean out your aquarium, your Axolotl can be gently cradled and lifted into temporary accommodation.**

## FOODS AND FEEDING

Axolotls are predators and they will eat a large variety of foods. Worms, crickets, small fish, fish pellets that sink or float, cubes of heart or lean meat are all suitable foodstuffs – mine are especially partial to small freshwater shrimp. Food pieces should be kept to small, mouth-sized chunks because Axolotls cannot chew and will never learn to use a knife and fork!

**▲ Requiring only a few feeds per week, the Axolotl is a popular and low maintenance aquatic pet.**

A young Axolotl needs 'training' to learn how to catch non-moving food, and long tweezers can be used to drop a piece of food onto its face. They suck in their prey and this violent motion can scare you from using your own fingers, but their teeth are very fine and will not hurt the adventurous keeper willing to hold a piece of food between finger and thumb. Axolotls can be fed every other day, with several mouthfuls of food given per sitting. Hungry Axolotls do seem to patrol their environment more than their well-fed companions, an indication that more food may be required. Uneaten food should be removed as soon as possible to minimize fouling.

## BREEDING

Once your pet is 2 years old the sexes can be distinguished easily. Viewed from above, the male Axolotl's head is longer and narrower than the female's, his tail is longer and the swelling by the cloaca is greater. Sexing is easiest when several specimens can be viewed together.

In captivity, the breeding season coincides with our winter and spring, when a change of temperature often triggers breeding behaviour. This can be induced artificially by raising the water temperature to 22°C (72°F) for about a week or so, and then allowing it to drop quickly. With luck and a good diet, breeding behaviour should follow. The female becomes attracted to the male's sweet excretions produced from his cloaca. She follows him around the aquarium, and they dance and swim around one another for some time. Eventually the male releases triangular jelly masses, called spermatophores, that sink to the bottom of the aquarium.
The female is led over these masses until she takes some up into her cloacal opening. Some hours later she will start to spawn, and some 300-600 fertilized eggs are laid.

The egg masses are best reared separately from the parents to prevent them from being damaged or eaten. The eggs need good, but not too strong, aeration from a pump. After about two weeks at 20°C (68°F) the larvae will hatch and can then be fed on brine shrimp, tubiflex, powdered fish food, or other microscopic foods. As they grow, daphnia, glassworms, bloodworms and mosquito larvae can also be consumed. Ample space is required to ensure a good survival rate for the hundreds of young, otherwise they will constantly snap and bite at each other.

# AFRICAN CLAWED FROG

*Xenopus laevis*

**These amphibians are totally aquatic and are noted for the absence of both tongue and eyelids. Their bodies have a somewhat flattened appearance and they possess powerful hind limbs.**

Native to southern Africa, these clawed frogs grow to approximately 10-13cm (4-5in) and in captivity they can live to 15 years of age. They are powerful and agile swimmers with fully-webbed hind limbs that are partially clawed. They are a simple species to keep, making them a popular choice for the novice keeper.

## CREATING THE RIGHT ENVIRONMENT

Simple to maintain in captivity, these frogs have been bred for many generations under laboratory conditions. A 90 x 38 x 30cm (36 x 15 x 12in) aquarium, filled with water to a depth of about 13-20cm (5-8in) is adequate for a similar sized pair of frogs. To check for correct water depth ensure that your frogs can reach the surface of the water to breathe while standing on the bottom of the tank. Keep water temperature at 20-25°C (68-77°F), using an aquarium heater if necessary, making sure that it is adequately protected from any possible damage by the frogs. To maintain water purity, gentle filtration using internal or external power filters is essential.

▲ **Simple to keep, the African Clawed Frog is a popular if unusual looking amphibian.**

To furnish your setup, use plants, rocks, pea-sized gravel and you can even add flower pots for your frogs to hide in. Supplementary lighting is not required. Your frogs may take a little while to settle into their new home. If they frantically bash themselves against the sides of the tank when you approach, insert a dense, plastic plant at one end of the aquarium and allow the frogs to hide and gradually acclimatize to their new surroundings.

## Dwarf Clawed Frog

*Hymenochirus boettgeri*

In the wild, these frogs from west Africa can be found in ditches and lakes at the edge of forests or cultivated areas. Measuring only 3.5cm (1.5in), they prefer an aquarium heated to 20-25°C (68-77°F) with a platform of floating plants to provide easy access to the water's surface. These small frogs breathe air, so access to the surface is essential. Suitable foods include daphnia, tubiflex and small insects such as hatchling crickets and fruit flies.

### HANDLING

These frogs are very difficult to handle. Coated with a protective slime, they slip through fingers even better than wet soap. If you need to move them, catch them in a net and transport them in plastic containers, without water, furnished with wet moss or wet foam pieces to keep your frog moist. Beware of transporting an African Clawed Frog in a bag of water because their claws may actually puncture the plastic.

### FOODS AND FEEDING

A healthy African Clawed Frog will rarely refuse food, which it shuffles about with its long fingers. It will happily swim to the surface and take food from your fingers, which it them consumes underwater. Earthworms, small fish (e.g. goldfish, minnow or guppy), waxworms, turtle and fish pellets are all agreeable foods to these frogs. Like many amphibians they are easily overfed because they just don't know when to stop – in the wild they would never find a constant and abundant food source all the year round. Feed only as much as they will eat per 10-minute sitting, 3-4 times a week.

### BREEDING

Males and females are easy to tell apart. The female is larger and more rounded and, when viewed from above, three flaps of skin next to the cloaca are visible.

Some environmental change seems to encourage spawning. Spraying the water surface or adding 5-8cm (2-3in) of water can start the process. Sometimes, however, the frogs spawn without any encouragement. On average, 1,000 eggs are laid per spawning and a single female can produce up to 10,000 per season. These should be removed immediately, to prevent the adults from eating them, and be kept in a shallow container until they hatch a couple of days later. The tadpoles have cannibalistic tendencies, so thin them out constantly, isolating faster growing specimens.

The organic soup required to feed the larvae may be purchased as a fish fry food, or by micro-mesh net dipping in a 'good' pond (a chemical-free freshwater environment containing an abundance of micro-organisms).

# ORIENTAL FIRE-BELLIED TOAD

*Bombina orientalis*

**These hardy toads are ideal for the hobbyist. They are small, easy to keep, very active and beautifully marked – easily one of the world's most attractive amphibians.**

Native to cool mountain streams and rice paddies in China, Korea and Siberia, these little toads can grow to about 5cm (2in) in length. Oriental Fire-bellied Toads have a bright green back with black markings, while their underparts are bright orange with black patches.

## CREATING THE RIGHT ENVIRONMENT

Two options exist for keeping these toads in captivity. A large aquarium – 90 x 38 x 30cm (36 x 15 x 12in) – which is ample space for three to four individuals (one male to three females); or, a moist woodland setting. In the former, the water should be maintained at a depth of 15-20cm (6-8in) and heated by a thermostatically-controlled aquarium heater to a temperature of 23-25°C (74-77°F). Natural daylight is sufficient but a small spotlight or fluorescent tube may be used for aesthetic purposes. Real or plastic plants and gravel will enhance the display, and rocks or cork logs should be used to allow the frogs to climb out of the water.

▲ **The small Oriental Fire-bellied Toad makes an excellent choice for a novice pet keeper.**

Make sure the tank is securely covered but well ventilated – these toads are excellent escapees given the opportunity. An internal water filter should ease maintenance, but larger faecal pellets and uneaten or dead food are best removed with a siphon. The water should be partially changed every few weeks. Remember to wash all furnishings and gravel at the same time.

Alternatively, keep these toads in a moist woodland-style setup furnished only with mosses, logs and a large, shallow water dish and maintain at 23-25°C (74-77°F). They will only need to be placed in a greater volume of water for breeding purposes.

**HANDLING**

Handling a happy toad is safe but not recommended regularly, nor is it even necessary to enjoy it. Like other amphibians, these toads contain known toxins and you should not handle them if you have cuts on your skin. When threatened, they lie flat and curl all their fingers and toes above their backs. This shows any predator that they are brightly coloured and potentially toxic. If actually attacked, an acrid, white fluid can ooze from pores on the back – this is toxic if swallowed. The threat display and release of toxins is rarely seen and should not deter you from acquiring this species.

**FOODS AND FEEDING**

These little toads feed well, and prefer live insect prey – crickets, locust hoppers, spiders, flies, etc. They feed mainly on land and will soon learn to come to your fingers when offered food. Feed as much as will be eaten 2-3 times a week and add vitamin supplements regularly. The superb colouration of Oriental Fire-bellied Toads will fade with captive-bred offspring unless colouring agents are provided (found in prepared fish foods and flakes). The toads obtain these naturally by eating a multitude of tiny wild foods.

**BREEDING**

Spring is the most likely time for these toads to start calling. The male is particularly noisy and his constant 'croaking' should attract a willing female. Spawning will occur within 8-24 hours of mating, probably during the night. Up to 200 eggs may be laid per spawning, singularly or in clumps. Eggs should be transferred to a rearing container filled with similarly aged and warmed water. The 2mm (0.08 in) eggs will hatch within a few days.

Once the larvae have absorbed their yolk sacs, they will be ready to start feeding. Finely chopped tubiflex, hatchling crickets and aquarium fry-food can all be given. Remove waste matter regularly. About 3 weeks after hatching the toads are ready to leave the water for the first time. They will be mature enough to breed at about 12 months.

## Giant Toad

*Bufo marinus*

This is the largest known species of toad, measuring 25cm (10in). Native to South and Central America and the Caribbean, these toads have, however, been foolishly introduced to Australia by man to eat a plant pest called the cane beetle. They are ravenous feeders and will eat almost any animal they can fit into their mouths. This terrestrial species can become very tame. They require a large vivarium – 90 x 38 x 30cm (36 x 15 x 12in) – with temperatures of 20-25°C (68-77°F). This giant toad is also a giant breeder, laying some 20,000-30,000 eggs per spawning.

# FIRE SALAMANDER

*Salamandra salamandra*

**These secretive creatures are mainly nocturnal, spending the day hidden under loose rocks or inside old rodent burrows. A cool and moist, mossy forest floor is their ideal habitat.**

Highly variable in patterning, the bright yellow colouration of the Fire Salamander is very striking and you are highly unlikely to mistake it for any other creature.

Their natural range extends through central and southern Europe into North Africa and western Asia. They are easy to keep in captivity and can grow to 30cm (12in) in length.

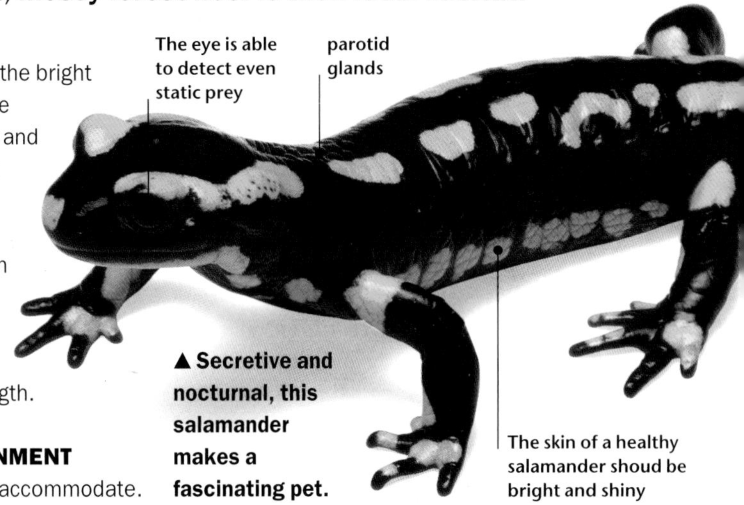

The eye is able to detect even static prey

parotid glands

The skin of a healthy salamander shoud be bright and shiny

**▲ Secretive and nocturnal, this salamander makes a fascinating pet.**

## CREATING THE RIGHT ENVIRONMENT

Fire Salamanders are very easy to accommodate. However, they are quite territorial and it may be easier to house individuals separately. A vivarium measuring 90 x 38 x 30cm (36 x 15 x 12in) or 120 x 38 x 30cm (48 x 15 x 12in) provides spacious living quarters and is ideal for a single specimen or a pair. If you do decide to keep a small group, provide them with plenty of hiding places and adequate food to minimize competition.

Furnish your vivarium with several centimetres of peat, moss and leaves to provide a cool, damp base, and add some cork or a small rotting log to provide hiding places. A shallow dish sunk into the ground is useful; salamanders drown very easily, so keep the water shallow and use pebbles or a slope to ensure the salamander can easily climb out. Mist spray your pets regularly.

Fire Salamanders, like some other amphibians, are visually aware and set up 'home' at a regular site. When cleaning out vivaria, it helps to replace some of the 'landmark' objects in their original spots to prevent confusing your pet.

These amphibians like cool, low-light conditions, 15-20°C (60-68°F), so heating is not usually required, but use a thermometer as a guide. Salamanders will look for suitable habitats if necessary. During my teens my Fire Salamander escaped. Over a year later my father was re-plumbing the bathroom and it re-emerged, giving him quite a shock. It had found suitably damp conditions and probably survived on house spiders, wood lice and stray worms.

## HANDLING

The bright colours of the salamander serve as a warning to any potential predator. Many creatures learn very quickly that these rubbery amphibians ooze toxic, milky secretions from their skin when bitten or aggravated. Therefore great care should be taken when handling your pet salamander. Personally I have never heard of any unhappy incidents involving salamanders, but it is best to be aware of this unpleasant defence mechanism. Enjoy your salamander but, as with all animals, treat it with respect.

## FOODS AND FEEDING

If fed at a regular time, your salamander will learn to emerge from hiding in anticipation of food. These amphibians eat a range of live foods. As a terrestrial species climbing foods, such as locusts, are best ignored. Slugs, spiders and earthworms, which remain at the salamander's level, are by far the most suitable foodstuffs.

Be very careful when feeding several salamanders at the same time because they are more than likely to fight over the same bit of food. As a simple rule, feed your pet salamander as much as will be eaten at a single sitting, 2-3 times per week. Without distressing your salamanders too much by leaving excess live food crawling all over them, leave a few worms and insects loose in the vivarium to be eaten or found at their leisure.

## BREEDING

Fire Salamanders are frequently bred in captivity, usually during the spring and early summer. Courtship occurs on land and involves the male carrying the female around, often for several hours, on his back. After stimulation of her cloacal region the female picks up a deposited spermatophore. At this time the female needs access to shallow water.

She will eventually bear up to 70 well-formed larvae direct into the water – ensure that the water is filtered gently. These larvae have developed inside the female's body from eggs, and like Axolotl larvae, feed on tiny organisms in the water known as infusoria. The larvae should be transferred to a rearing container to prevent the adults eating their offspring. Some 3-5 months later metamorphosis will take place; at this stage the gills will reduce as the young start to breathe oxygen. In some populations the female develops the larvae for longer inside her body and gives birth to fully developed juveniles, already coloured with the yellow blotches and stripes.

### Tiger Salamander

*Ambystoma tigrinum*

These attractive salamanders are native to North America. Growing to 33cm (13in), they are the largest of the terrestrial salamanders. Colouring tends to be variable, ranging from black to brown, with bright spots or blotches of light grey. They tame easily and enjoy eating live foods such as worms and snails. Living conditions should be similar to those of the Fire Salamander, with an ideal temperature range of 15-23°C (60-74°F).

# RED-SPOTTED NEWT

*Notophthalmus viridescens*

**Native to eastern North America, these charming and attractive newts make an ideal introduction to the world of tailed amphibians.**

Red-spotted Newts go through a 'red eft' stage upon metamorphosis. These juveniles are usually red with black dots on their backs and lead a terrestrial lifestyle. Once adult, their skin becomes brown and is covered in red spots, encircled with a black border. At this stage, they return to the water to live and breed. Easy to maintain, these newts grow to a length of 10cm (4in) and live for 4-6 years.

## CREATING THE RIGHT ENVIRONMENT

The habitat requirements of these newts change as they mature. Metamorphosed juveniles are terrestrial but adults are mostly aquatic. The 'red efts' take 2-3 years to mature and only then will they return permanently to the water as adults. Several newts may be accommodated in a 60 x 38 x 30cm (24 x 15 x 12in) aquarium, but one measuring 90 x 38 x 30cm (36 x 15 x12in) would give you more scope to create a semi-aquatic setup to accommodate both the aquatic and terrestrial stages of a newt's life cycle.

A well-planned vivarium consists of a densely-planted aquatic region, with a water depth of 15-23cm (6-9in), and a large platform, or cork logs wedged or stuck to the inside of the tank, to create a land area. Sloping branches or rocks provide easy access to the water.

Cover the land area with leaf litter, wood chips, moss and pieces of log or cork for your juvenile to utilize as living and hunting spaces. They do not appreciate damp and stale conditions or habitats that are too dry, so humidity should be moderate, say around 40-50%. Full-spectrum lighting may enhance the look of your setup but it is not vital for newt health – good natural daylight is sufficient, however, do not place in direct sunlight.

Mucus glands are located all over the body and produce a slimy secretion to protect the skin and keep it moist

Supported by a very delicate skeleton, the tiny body makes it an unsuitable pet for regular handling

The tail makes up half of the overall body length

**▲ Red-spotted Newts are agile and fast hunters of the water's edge.**

## Great Crested Newt

*Triturus cristatus*

These newts are a native British species, growing to approximately 17cm (6.5in) in length. Their numbers are declining in the UK due to habitat loss, so they are protected by law and cannot be collected from the wild or have their habitats destroyed. They must only be purchased from captive-bred stocks and it is good to try to keep and support, even breed, these native amphibians.

Red-spotted Newts are content to live at room temperature, 18°C (65°F). The environment should not become too hot and the maximum recommended temperature is 22°C (72°F). Use an external power filter for optimum water quality. Add plants to the water to provide an attractive environment for your newts. Plastic plants are more hygienic because they do not rot. If you wish to use real plants, use fluorescent lighting to encourage good plant growth.

▲ **These newts will patrol the vivarium searching for small insects, looking both above and below the water's surface.**

### HANDLING

These newts produce toxic skin secretions, making them unpalatable to many predators. Although these toxins will not make you ill, it is best to handle your newt as little as possible. If you do need to handle them, do take care since they are small, which means they must be handled with care to avoid an injury.

### FOODS AND FEEDING

Terrestrial efts will hunt small live foods such as young crickets, waxmoth larvae, fruit flies, worms, spiders and beetles. Aquatic adults hunt amongst the plants and search out daphnia, mosquito larvae, bloodworm and other aquatic micro-organisms, many of which can be purchased at pet stores. Smaller live foods (i.e. crickets and fruit flies) may also be dropped onto the water surface. Offer food on a regular basis and leave some live foods in the vivarium for your newts to hunt.

### BREEDING

Males actively court the females in water during springtime. Invisible scents, called pheromones, are produced by the male's cloacal glands and these are directed toward a potential mate using the tail. Eventually, a receptive female will pick up the spermatophore directly into her cloacal region. A few weeks later she will lay up to 400 individual eggs. These hatch in about 4-8 weeks and the tiny larvae begin to feed on suspended micro-organisms for about 3 months. During this time they will have grown from 1cm (0.5in) to 4.5cm (2in) in length and will be ready to undergo metamorphosis.

# TREEFROGS

*Family: Hylidae*

**Tree frogs are the acrobats of the amphibian world. Possessing comical swollen pads on their digits, these lively creatures can reach parts of the vivarium other frogs can only dream about.**

Found in warmer climates across the globe, treefrogs are often extremely colourful and several species are successfully kept and bred by herpetologists. As their name suggests, the majority of treefrogs are arboreal, spending most of their time off the ground in plants and trees. Any vertical surface may be climbed, even glass. Having found a suitable spot, they will curl their forelimbs and legs close in to the body and rest, tucked away like this all day, awaiting night fall or more crepuscular conditions.

## WHITE'S TREEFROG

*Litoria caerulea*

Large, chubby and with a permanent smiling expression, this species is by far the most common pet treefrog. Native to north east Australia and New Guinea, these arboreal frogs live in rainforest-type habitats. When adult, this species can reach a length of 5-10cm (2-4in) and can live for up to 15 years. Their colour can vary from dark olive to vibrant green.

### CREATING THE RIGHT ENVIRONMENT

Tall vivaria are required for most treefrogs and White's are no exception. A vivarium measuring 60 x 45 x 30cm (24 x 18 x 12 in) and filled with robust branches, plants and furnishings is ideal for one or two adults. A substrate of moss and bark chips, stones and pebbles is recommended. A large bowl of water should be offered and daily mist spraying is beneficial. However, make sure there is a reasonable amount of ventilation to prevent excess dampness. Keep the temperature between 25-30°C (77-86°F) during the day and reduce slightly at night. If you plan to keep more than one treefrog, make sure that individuals are of the same size to prevent cannibalism.

### HANDLING

Whilst individual characters vary, most White's tolerate handling well. Ensure that your hands are moist to prevent any damage to their delicate skin. Sometimes, when your frog is nervous or exercising, a small cupful of liquid can be expelled from the anus. This liquid is often assumed to be urine, but it is only water.

**▲ This is one of the easiest species of treefrog to handle. The large toe pads feel sticky and help this species to grip your hand securely.**

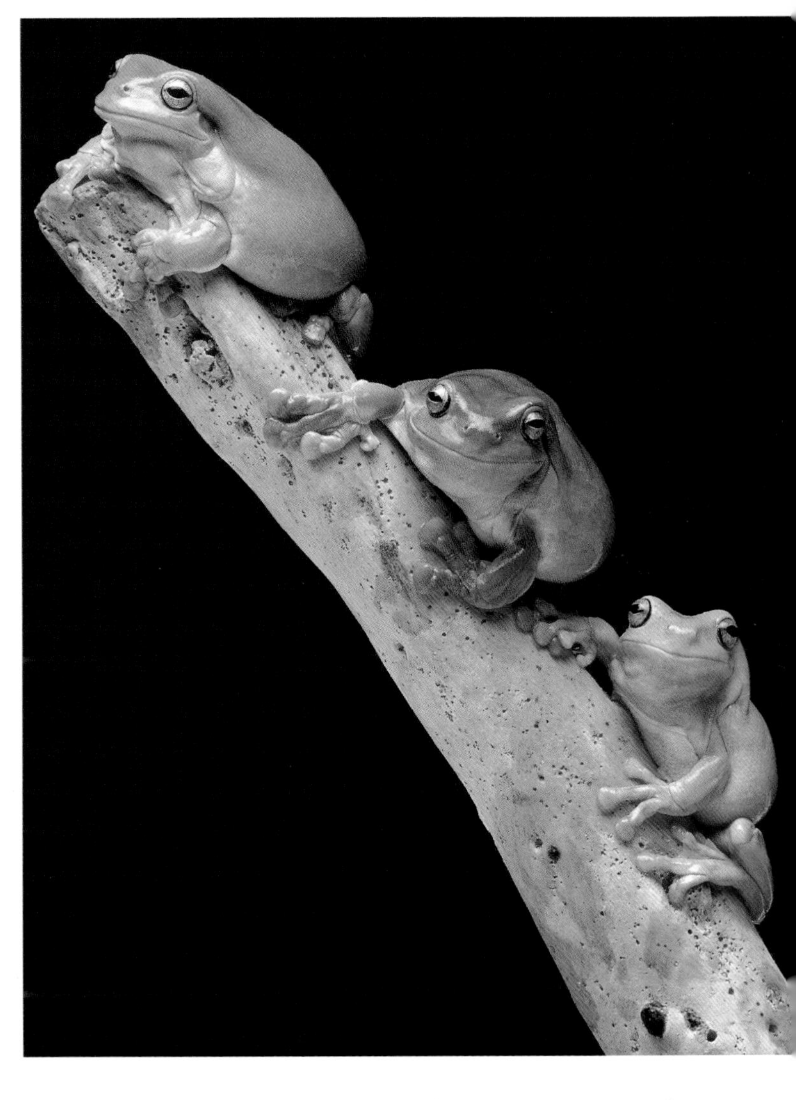

▶ **This trio of White's Treefrogs show a range of colouration. Possessing a fixed smiling expression, this greedy species grows rapidly and requires sturdy furnishings to support its weight.**

## FOODS AND FEEDING

These frogs are quite capable of eating small dead vertebrates, such as other frogs, lizards and pink mice, but they will be just as happy to eat live foods such as large locust hoppers, big crickets and earthworms. Depending on the size of your pet, feed 3-5 insects to each frog per sitting, 2-3 times per week.

## BREEDING

A separate aquarium measuring 60 x 38 x 30cm (24 x 15 x 12in), with a water depth of several centimetres at 25°C (77°F), is needed to attempt breeding. Once your frogs are in their new environment a simulation of approaching rains will encourage mating. Try reducing humidity to 40-60% and the temperature to 18-20°C (65-68°F) for a week or so. Then gradually raise the humidity to 90% and the temperature to 25-30°C (77-86°F) and spray frequently with water to imitate rainfall.

It is essential that the frogs can easily enter and leave the pool at will. A sturdy branch or rock positioned against the edge of the water area will be sufficient. Between 200-300 eggs are laid as spawn and these will hatch in approximately 24 hours. Within 4-6 weeks the larvae, or tadpoles, will have grown sufficiently to trigger the process of metamorphosis. The young frogs will grow rapidly and should be separated into smaller groups to prevent overcrowding. They will mature between 1-2 years of age.

## AMERICAN GREEN TREEFROG

*Hyla cinerea*

In the wild this species is found in the forests and meadows of southern North America. Their bright green bodies are quite slender and they have a cream stripe running down each side of the body. Measuring only 4-6cm (1.5-2.5in), they are ideal for keeping in a vivarium; indeed, they are one of the most popular types of treefrog available to the hobbyist.

### CREATING THE RIGHT ENVIRONMENT

A semi-moist vivarium suits these treefrogs. Height is essential and three or four specimens may be housed in a vivarium measuring 60 x 46 x 30cm (24 x 18 x 12in). During spring and summer, maintain at a temperature of 18-25°C (65-77°F), cooling to 12-18°C (54-65°F) at night. Cover the land area with mosses, wood chips, twigs, and leaves and, even out of breeding season, a water dish should be available at all times. Ideally, the ground area should be divided into three parts land and one part water.

Insert branches or cork logs to provide climbing opportunities. Ferns or robust Philodendron plants should also be considered.

Avoid using unprotected incandescent light or spot lights because the light will not encourage activity and treefrogs often injure themselves fatally when leaping onto a hot light bulb. Heat pads and protected heaters are fine. A full-spectrum light tube should be provided since treefrogs benefit from exposure to UV rays.

### HANDLING

This species is quite easy to handle but, again, ensure that your hands are wet. American Greens tame relatively quickly as they respond well to feeding from the hand.

### FOODS AND FEEDING

Many treefrogs prefer flying foods; blue bottles and house flies excite them into activity and provide much amusement for the observer. Most live foods, such as crickets, flies, locust hoppers and worms will be accepted. Feed your American Greens 2-3 times per week; as a general rule they eat more in spring and summer than in winter.

### BREEDING

American Green Treefrogs are most likely to breed after a period of over-wintering or hibernation, followed by re-housing to larger quarters supplied with greater amounts of water. The short breeding season occurs around April and May, and the males are very loud when calling. After amplexus, the coupling pair enter the water. The eggs are fertilized by the male as they

### European Green Treefrog

*Hyla arborea*

The European species is fairly rounded and has a brown stripe running through the eye and along

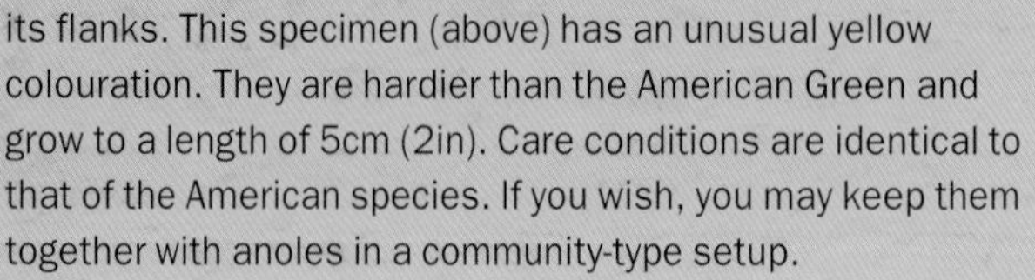

its flanks. This specimen (above) has an unusual yellow colouration. They are hardier than the American Green and grow to a length of 5cm (2in). Care conditions are identical to that of the American species. If you wish, you may keep them together with anoles in a community-type setup.

▲ **This American Green Treefrog is resting on a piece of cork bark ready to leap at prey.**

leave the female's body and several clumps of 100-300 eggs are deposited amongst the water plants. Transfer these eggs to shallow rearing containers where the water is filtered, aerated and kept at about 20°C (68°F).

Several days later the larvae, or tadpoles, hatch and should be fed on microscopic organisms and fine powdered fish food. Metamorphosis takes place about a month later and the froglets, measuring 1-2cm (0.5-0.75in) and brown or green in colour, are ready for a more terrestrial life. They mature at 2-3 years.

## AMAZING FACT

Most temperate amphibians hibernate to avoid lethally cold winters. Upon emerging in spring, most species will be eager to breed. Replicate this rest period by cooling your frogs gradually to 5-10°C (41-50°F). You will find that even just a few weeks at reduced temperatures will encourage your pet to breed once they are warmed up again. Before trying this for yourself, ask a specialist dealer for more advice.

▲ Less hardy than other treefrogs, the Red-eye is a popular and visually stunning amphibian pet.

## RED-EYED TREEFROG

*Agalychnis callidryas*

One of the most striking of all amphibians, this treefrog has bright red eyes, orange feet and blue and white stripes on its flanks. They are frequently available from captive-bred stocks and are best purchased when approximately 1-2cm (0.5-1in) in length, growing to 8cm (3in) when adult. Starting with young healthy specimens seems the best approach because they should settle in fairly quickly within the conditions you are providing.

### CREATING THE RIGHT ENVIRONMENT

Tall vivaria, measuring 45 x 60 x 30cm (18 x 24 x 12in), furnished with real plants and branches should be provided for these wonderful frogs. Attaining the correct balance of humidity, temperature and ventilation is important for the successful care of these and, indeed, most vivarium pets. These creatures prefer 50-80% humidity and a daytime temperature of 25-30°C (77-86°F), dropping to approximately 20°C (68°F) at night when they become active. Make sure ventilation is adequate.

A water bowl must be available at all times and should be cleaned every other day because these, and other amphibians, absorb water into the bladder – dirty water may often lead to fatal infections. A full-spectrum light source benefits most species and also encourages good plant growth.

### HANDLING

These bright frogs tend to be slow moving and ponderous by nature. They are nocturnal and when at rest during the day the bright colours are completely concealed from view by the foliage. Handle cautiously, and only when necessary, because their skin is extremely delicate – it is far better to enjoy this species visually. If you need to move them, wait until they are at rest on a leaf and then place them in a small tub with air holes. Keep them warm and moist.

### FOODS AND FEEDING

Red-eyed Treefrogs are particularly fond of flying foods such as flies and moths, but crickets and other live foods of suitable size can also be offered. Dust the food with multi-vitamins and give each frog as much as will be eaten at each sitting, 2-3 times per week.

### BREEDING

In the wild, these arboreal frogs breed in bushes, shrubs and trees that overhang water. Eggs are laid in a frothy mass on leaves and the tadpoles, on feeling rain, drop from the leaf into the water below. To trigger breeding in captivity, try the simulation technique used for the White's Treefrog. A large bowl or small pool of water will be necessary. Tiny froglets require small insect foods dusted with supplements to ensure healthy development.

**▲ Small and delicate amphibians require careful handling. Red-eyed Treefrogs are best left in their vivaria and handled only when necessary.**

# Reptiles 1

## Lizards and Tortoises

Reptiles are primitive vertebrates that have been around for millions of years. Their waterproof, scaly skin enables them to thrive in a variety of habitats. This group of animals includes lizards and tortoises.

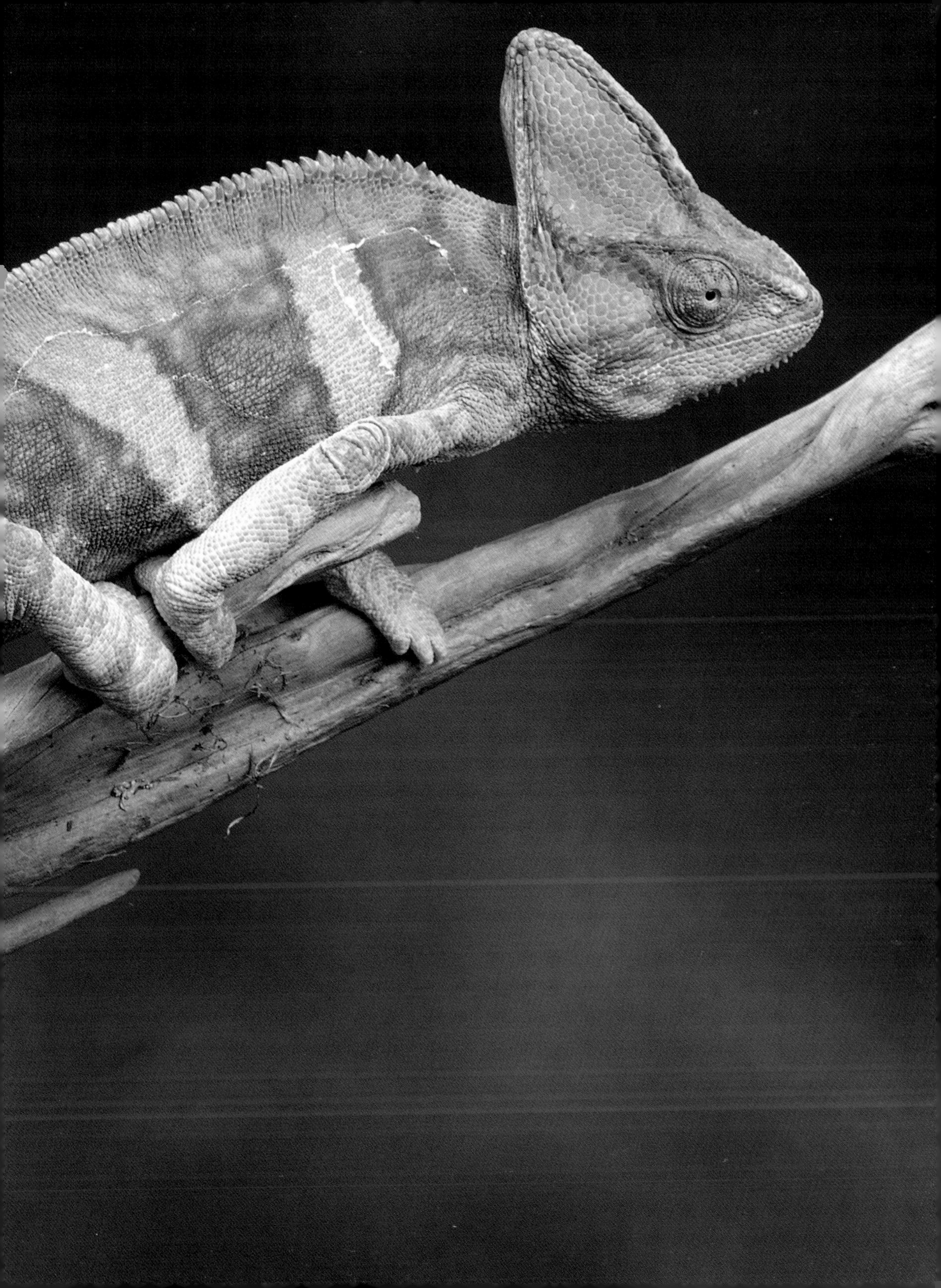

The term herpetology, the study of reptiles and amphibians, is derived from the Greek word 'herpeton', which means a creeping animal. Indeed, both lizards and tortoises do crawl, but the extent of this type of behaviour is essentially controlled by their need to thermoregulate.

Unlike mammals, which utilize food to maintain a steady internal temperature, the temperature of a reptile is entirely dependent on its external environment. Reptiles need to bask in the sun or sit upon warm ground when they are cold and, conversely, they must retreat into shaded areas if they are too hot to avoid fatal overheating. The ability to darken or lighten their skin also helps this process – darker skin absorbs heat, while lighter skin reflects it. Reptiles are often misleadingly referred to as 'cold-blooded' animals, but during basking their blood may reach temperatures that are warmer than that of a human.

During hot and cold seasons, when temperatures reach extreme levels, reptiles will become inactive. Aestivation and hibernation are the terms used to describe this process, whereby the metabolism is greatly reduced and reptiles enter a torpid, inactive state. Aestivation occurs in tropical areas and hibernation takes place in temperate regions.

Reptile skin is effectively watertight because it is covered by protective scales. In some species, like the chelonia, these scales are enlarged and contain bony plates which, along with flattened and fused vertebrae, form the shell. Scales are formed from an outer layer of dead skin; in order for the enclosed animal to grow, this layer is shed or sloughed off regularly and renewed from underneath. Many lizards eat this recyclable material. The skin of reptiles may appear to be shiny but it is never slimy. Reptiles do not sweat and, unlike mammals, they do not have sebaceous glands.

Most species eat a variety of foodstuffs but generally they consume more animal matter than plant material. Several lizards have even developed unique predatory techniques to lure prey. For example, the long tongue of the chameleon and the insect-luring tail of many species of skink.

The majority of lizards, and all chelonia, are egg layers. The eggs may be deposited into leaf litter, buried in the ground or even stuck to tree bark. Some eggs have a hard shell while others have a leathery, pliable composition. Once eggs are laid, they are incubated. Reptile young are always born, or hatch, as miniature replicas of their parents – no metamorphic stage exists as it does in amphibians.

**▲ Tortoises are long- lived reptiles, with some living to over 100 years of age.**

◀ Thousands of lizards and tortoises are bred in captivity for the pet trade, greatly improving the quality and suitability of specimens for the enthusiast.

## REPTILE CLASSIFICATION

| Order | Description | No. of Species |
|---|---|---|
| SQUAMATA | Lizards and snakes. This group of reptiles is split into two sub-orders, the Sauria (lizards) and Serpentes (snakes). Lizards usually possess four legs, a long tail and moving eyelids, while snakes are limbless and have fused eyelids. | 5,150 |
| CHELONIA | Tortoises and turtles. These creatures possess a bony shell made of horny plates which enclose their soft body. Terrestrial and aquatic forms exist. | 230 |
| CROCODILIA | Alligators, crocodiles and gharials. These are large predatory reptiles which live in or near water. If kept in captivity, crocodilians must be licensed under the Dangerous Wild Animals Act. | 21 |
| RHYNCHOCEPHALIA | Tuatara. This lizard-like reptile evolved before the dinosaurs and is the sole survivor of its order. It is an endangered species found only on a few islands off New Zealand. | 1 |

# LEOPARD GECKO

*Eublepharis macularius*

**These velvet-skinned lizards are marked as their name suggests, like a leopard, and are found in the harsh, arid habitats of north-western India and Pakistan.**

Leopard Geckos are members of the Gekkonidae family. Most members of this family are noted for their climbing abilities. Fine hair-like structures of specially modified scales enable them to climb glass or walk across a ceiling. Leopard Geckos, however, are a ground-dwelling species, which use their sharp claws for digging and for clambering about on rocks and logs. Fully grown adults measure 20-22cm (8-9in).The young, thousands of which are bred in the UK for pet keepers, are generally purchased when they are a few months old and 7-10cm (3-4in) in length. These geckos can live for up to 20 years in captivity.

**▲ The subtle differences of pattern and colouring in a reptile population are called phases – the light and dark phases for this gecko are shown here.**

In their natural habitat, different Leopard Gecko populations show some variety in colour strength and patterning. These subtle differences continue to be present in the captive-bred population and two phases are commonly seen. A third phase – showing a chocolate colouration – has been reported by herpetologists in North America, but I have not seen any in the UK.

## CREATING THE RIGHT ENVIRONMENT

The Leopard Gecko's natural habitat is arid and desert-like. To simulate this environment and to provide all the essentials for survival, your set-up needs to include the following: several shelters in which to sleep, hunt and remain unobserved; a shallow water dish for drinking; and an area that can be sprayed regularly to provide a humid micro-climate. The latter ensures that your lizard does not dehydrate and helps to facilitate periodic skin shedding.

Leopard Geckos will thrive at temperatures between 27-32°C (80-90°F). A well-designed vivarium should provide your gecko with a hot spot and various cooler areas within this range to allow it to thermoregulate effectively. Use a heater pad at one end of the vivarium. The temperature may drop to 21°C (70°F) at night without any ill effect. These geckos are nocturnal, so no overhead light bulbs or ultra-violet light is necessary.

Play sand, wood chips, rounded gravel and pebbles can all be used as a base medium or substrate. Cork bark, cactus skeleton or other suitable material can be stacked or glued together to provide shelters. Live succulent plants, except cactus, may be introduced to good effect – simply bury the pot into the substrate.

Geckos are very clean and will choose one spot in their vivarium to use as a toilet. The dry droppings should be removed regularly. If housing a single specimen, your setup should be cleaned out completely once every 2-3 months – make sure the vivarium, rocks and plastic plants are also cleaned and disinfected. Your pet lizard must be housed in temporary accommodation whilst its home undergoes a thorough clean.

A 60 x 30 x 38 cm (24 x 12 x 15in) vivarium provides ample space for a single specimen, while a 90 x 30 x 38cm (36 x 12 x 15in) vivarium would be suitable for a trio. Professional breeders often keep adults, and rear juveniles, in considerably smaller containers.

Ideally, your Leopard Geckos should be housed separetely, where there is no competition for food or prime resting sites, and they will not be

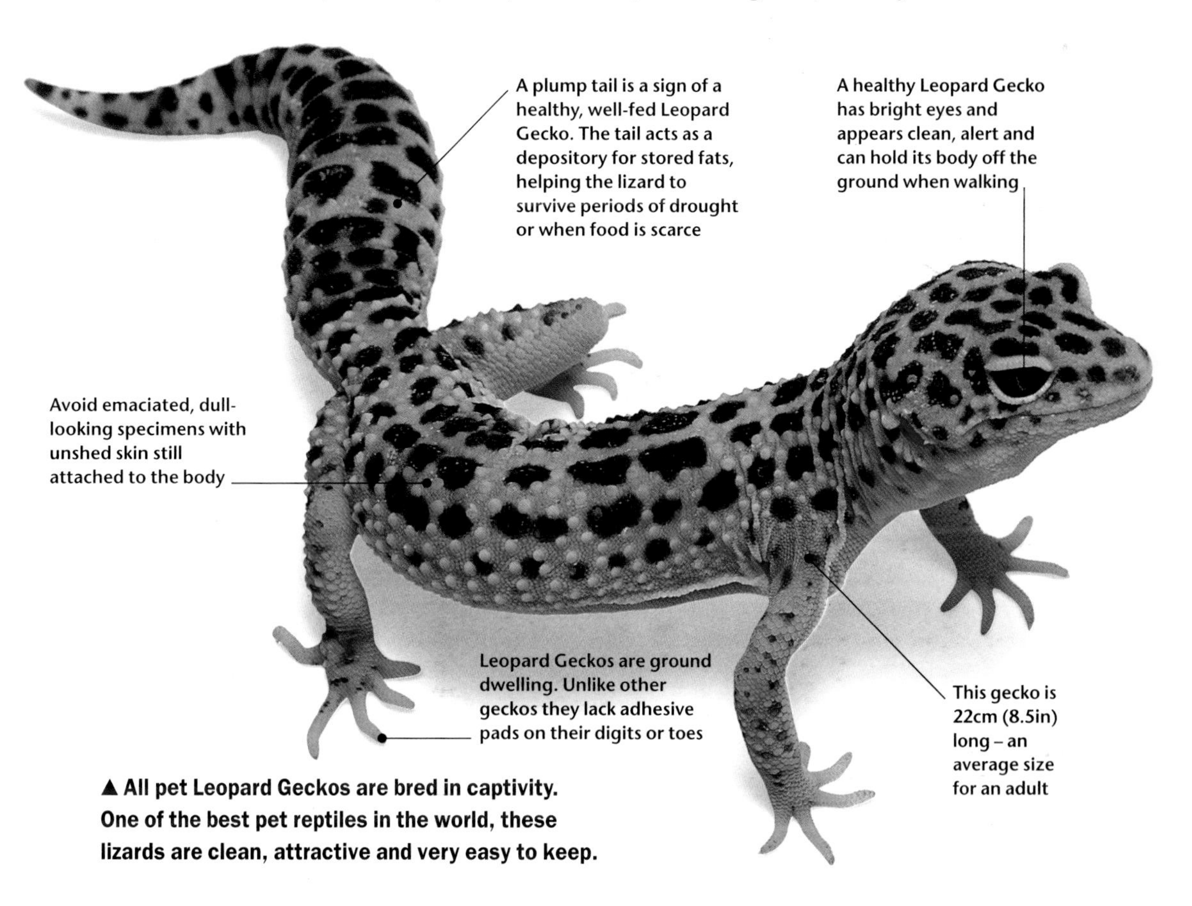


**▲ All pet Leopard Geckos are bred in captivity. One of the best pet reptiles in the world, these lizards are clean, attractive and very easy to keep.**

harassed by any other individual. It is very important not to keep males together because the strongest and most established specimens will bully smaller, weaker animals. Females can be kept together and a small colony of, say, four females and one male is also fine. One simple rule for colonies – the more space and hiding places you provide, the less any one individual will be harassed or become subordinate. Alternatively, you could keep a pair, housed separately, only introducing them for breeding for short periods, of a day or so, at a time.

### HANDLING

Leopard Geckos tolerate handling well but, like any pet, handling should be done with care, consideration and patience. Sitting down whilst holding your pet will ensure that there is less likelihood of injury if it jumps out of your hand or is dropped. The best time to start taming your lizard is when it is still young. Handle for short, regular periods to show the reptile that you pose no threat to its safety – offer some food by hand as a possible reward.

**▼ Unlike most geckos, this species has moveable eyelids that enable it to blink. This helps to keep the eye free of sand and dirt.**

**▲ A small plastic container is ideal as a hide box. Containing vermiculite, it is likely to be utilized by a female looking for an egg laying site.**

### FOODS AND FEEDING

Leopard Geckos should eat every other day or so. They may be fed most small to medium live foods, such as crickets, a few of which should always be available within the vivarium for the geckos to stalk. A young adult will eat, on average, 5-6 insects at a sitting, several times a week. Supply water in a very shallow dish, positioned level with the ground.

### BREEDING

Leopard Geckos breed well in captivity. You may wish to try this yourself, but make sure that you have a separate home ready for your new offspring. Mating usually occurs unseen in a covered area, and the first signs of a successful coupling is the female's noticeably swollen belly. Each female can lay several clutches of 2 eggs per year. She will try to find a damp and secure spot to lay and bury her eggs, probably under the shelter or amongst plastic plants heaped on the ground. Try to encourage her to lay her eggs in a suitable spot – an upturned plastic container, with a hole cut in one side, is ideal. Otherwise check the vivarium regularly, especially in the morning and evening.

The two eggs, being a standard clutch for most geckos, should be transferred to an incubator to prevent desiccation. Unlike many bird eggs, reptile eggs should not be turned – simply place in the incubating medium (I use vermiculite) and cover. Provide airflow but keep humidity at 70-80% and maintain a temperature of 30°C (86°F).

The incubating eggs should be checked on a regular basis – very shrivelled or mouldy ones should be removed to avoid contaminating healthy eggs. A white but slightly indented egg needs a higher humidity level. The young will hatch after 6-12 weeks. The juveniles are marked with yellow and black bands that will gradually break up into the familiar adult markings. They should be housed in small rearing containers, separately if possible. If kept communally, you must provide several hiding places and water dishes, removing the more dominant individuals once they begin to harass others. A daily misting will ensure good skin condition and will ease frequent shedding as the juveniles grow. Small crickets and waxworms are ideal first foods – geckos also adore small spiders. All foodstuffs should be lightly dusted with a multi-vitamin and mineral supplement.

## Day Geckos

*Phelsuma sp*

Like wonderful living jewels, these diurnal lizards look like they have just had a fight in a paint shop. Day Geckos are native to tropical Madagascar, Mauritius, the Seychelles and other islands in the Indian Ocean. The majority are arboreal and are 10 -15cm (4-6in) in length.

Day Geckos are ideal for a tall, preferably planted, humid vivarium and will remain active for most of the day. Accommodation measuring 60 x 45 x 30cm (24 x 18 x 12in) is suitable for 2-3 specimens. Full-spectrum fluorescent lighting is recommended for Day Geckos and to maintain a visually stunning, planted vivarium. If it is not practical to use live plants, geckos will live just as happily with plastic or silk plants. The temperature should be maintained at between 25-32°C (77-90°F) using a spotlight – humid, but not stagnant – with good ventilation.

As well as eating live foods, such as crickets, these reptiles enjoy licking fruit, nectar or sap from tropical plants. An upside-down jam jar lid, or bottle top, is ideal for serving honey, jam or small fruit pieces, such as mango or grape. They prefer drinking from droplets on leaves, so mist spraying once a day is beneficial.

### AMAZING FACT

The incubating temperature of many reptile eggs will dictate the sex of the hatchling. At 30°C (86°F) the sexes are likely to be mixed. At lower temperatures, 25-28°C (77-82°F), the majority of young are likely to be female. Raising the temperature to 32°C (90°F) should produce males.

# GREEN ANOLE

*Anolis carolinensis*

**Often called American Chameleons, these delightful green lizards are fascinating to watch – they are very active and regularly display their colourful throat fans.**

Native to the southern United States, Central America and the Caribbean, Green Anoles are one of the most commonly kept lizards in the world. They can often be found on walls, fences, bushes and shrubs in gardens and other man-made settlements.

These diurnal tree lizards are often called American Chameleons. They measure 13-20cm (5-8in) in length and the tail, essential for balancing, makes up two-thirds of their total length. An average life expectancy of 3-5 years is typical.

Males are very territorial and aggressive to one another. Much of their time is spent displaying ownership of a particular area, usually with much head bobbing and extension of their colourful throat fan, or dewlap.

## CREATING THE RIGHT ENVIRONMENT

These lizards need a tall vivarium with a moist environment, with an ideal daytime temperature range of 23-30°C (74-86°F), which may drop to 20°C (68°F) at night. At least one basking spotlight and a full spectrum tube are needed to assist the healthy development of these reptiles. The spotlight and light tube are also necessary to maintain healthy growth of the real plants that should be provided in a vivarium that is as tall and as large as possible.

For a harem of four females and one male, I would suggest a vivarium measuring at least 90 x 60 x 30cm (36 x 24 x 12in). At least two basking spots are needed and many suitable territories should be created. Provide some branches or logs and a suitable substrate, such as potting compost covered with a layer of woodchips or leaf litter. Males should be housed separately if a large vivarium is not available.

Cleaning will be largely aesthetic, mainly to remove visible droppings from leaves and the sides of the vivarium. Soiled areas of compost

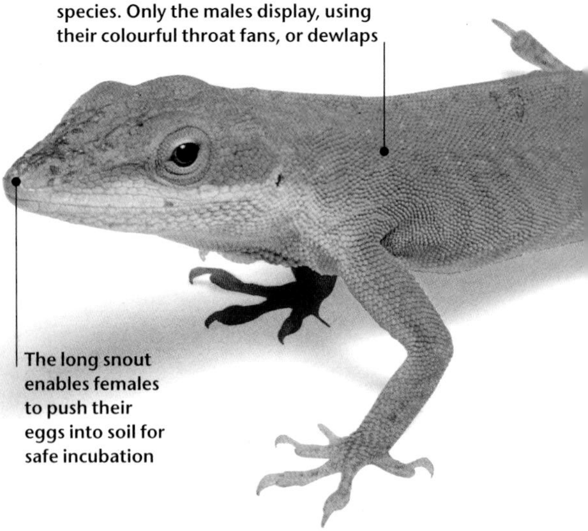

This adult male is more brightly coloured than the female of the species. Only the males display, using their colourful throat fans, or dewlaps

The long snout enables females to push their eggs into soil for safe incubation

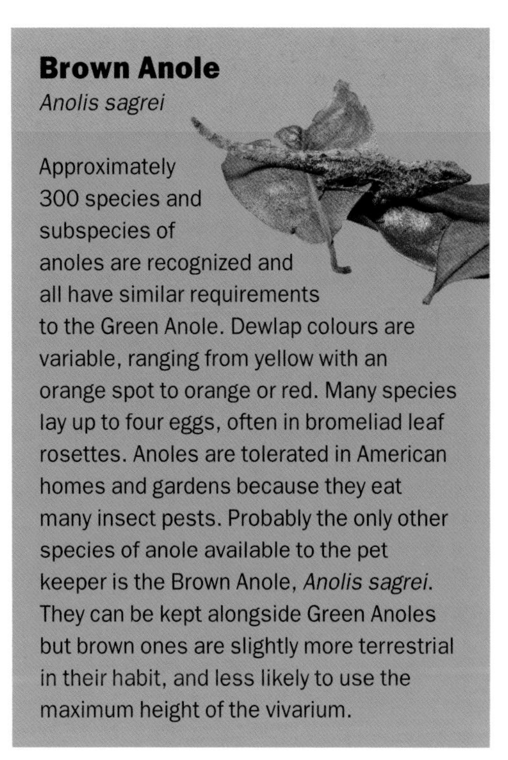

### Brown Anole
*Anolis sagrei*

Approximately 300 species and subspecies of anoles are recognized and all have similar requirements to the Green Anole. Dewlap colours are variable, ranging from yellow with an orange spot to orange or red. Many species lay up to four eggs, often in bromeliad leaf rosettes. Anoles are tolerated in American homes and gardens because they eat many insect pests. Probably the only other species of anole available to the pet keeper is the Brown Anole, *Anolis sagrei*. They can be kept alongside Green Anoles but brown ones are slightly more terrestrial in their habit, and less likely to use the maximum height of the vivarium.

The colour of these lizards can change from green to brown. Changes are the result of changes in mood, temperature, light intensity and stress as much as for camouflage purposes

**▲ Anoles are very active lizards. They can run fast on plant foliage and leap readily from one spot to another.**

can easily be scooped out. A planted vivarium will need tending like a garden rather than requiring the thorough cleaning necessary for a more artificial setup that uses, say, plastic plants and cork logs.

## HANDLING

Although fast and able to leap about, an anole will learn to sit on your hand in exchange for a tasty treat. These small and delicate lizards need a gentle touch. Allowing your anole to climb onto your hand is less stressful for the lizard, but be careful as it is likely to escape. If you wish to move your pet, grip it firmly but gently in the hand.

## FOODS AND FEEDING

These insectivores will eat any small to medium live insect foods. Mealworms should be avoided because many will pass through these reptiles undigested. Feed at least twice weekly by adding several live insects into the vivarium, watching to ensure that every occupant gets at least two insects per feeding. A small surplus amount of live food may be present in the vivarium at all times. You will soon familiarize yourself with when, and how often, your anoles need feeding by the size of their bellies and general behaviour.

Mist spraying once a day or so is essential because they prefer to drink water droplets from leaves rather than still water from a bowl. The vivarium should not become too wet, however, because the environment will become stagnant.

## BREEDING

Once you establish a group of these lizards and they are well fed and happy, breeding is sure to follow. One or two eggs are laid about a fortnight after mating. These will be buried by the female, or nosed into the compost in a warm and damp area, frequently around the pot or roots of a climbing plant, such as Philodendron.

# BEARDED DRAGONS

*Pogona vitticeps*

**Native to Australia, these heavy-bodied lizards are partly arboreal and enjoy basking on tree stumps and garden fences. They have appealing characters and are easy to keep in a vivarium.**

Growing to 40-60cm (16-24in), Bearded Dragons are a social species and interesting behaviour, such as courtship, recognition, threat and appeasement, may all be observed within the vivarium. Spiky, elongated scales, or 'beard', cover the huge throat pouch, and can be extended when a Bearded Dragon feels threatened, making an open mouth and inflated body an effective threat posture.

Temperatures should reach 28-40°C (82-104°F) at the hottest spot in the vivarium during the day, but can reduce to 20°C (68°F) during the night. For this diurnal species, 12 hours of daylight is sufficient and a dark rest period is essential for sleep. A large, clean water dish should be available at all times for bathing, and your Bearded Dragon will enjoy an occasional mist spraying.

## CREATING THE RIGHT ENVIRONMENT

A dry woodland, or desert-style vivarium measuring 76 x 51 x 76cm (30 x 20 x 30in) is suitable for an adult pair or trio of Bearded Dragons. Insert cork logs or rocks for climbing on or burrowing under and cover the vivarium floor with bark chips or sand. Add leaves and pine cones, or live succulents, to enhance your setup. If two or more Beardies are kept together, provide several basking areas to minimize competition. This will also help to ensure that the dominant, or alpha, male does not unduly harass other individuals.

The external ear opening is clearly visible on most lizards

Long claws enable the Bearded Dragon to climb boulders in the wild

As this young adult matures, his beard will darken and may become inflated when he is threatened or when displaying to a potential mate

**▶ A variety of social behaviours can be readily observed in a group of these medium-sized lizards.**

Larger lizards often have a higher metabolic rate and they create large quantities of faecal matter. Soiled areas need regular cleaning and the water should be changed frequently.

Juveniles are normally acquired when they are a couple of months old – usually measuring about 15-17.5cm (6-7in). Two Beardies would require at least a 60 x 30 x 30cm (24 x 12 x 12in) vivarium for the first few months of their life. Heater pads are useful to ensure the substrate is dry and warm at all times. Full-spectrum, fluorescent lighting is essential for healthy growth and development – occasional exposure to sunlight is beneficial, but ensure the lizard does not escape its enclosure.

**HANDLING**

Bearded Dragons tame very quickly while they are still young, especially if tempted onto your hand with a treat. The claws may feel sharp on a young person's hand, but in general these lizards are very easy to handle. The best way to proceed is to pick up your Beardie gently and place it onto your hand.

**FOODS AND FEEDING**

Juveniles grow very fast and require a well-supplemented omnivorous diet. Smaller insects, pinkies and spiders are particularly relished. You can cool your bugs to make them less active and easier for your pet to catch. Chopped fruit and leaves should also be supplied. Juveniles should be fed daily, but be careful not to overfeed. Larger specimens and adults should be fed 4-5 times a week. A bowl of cuttlefish, cut into pieces or grated, allows individuals, especially females, to obtain extra calcium. This is particularly important for healthy egg development in gravid females. Regular vitamin and mineral supplements are required for optimum long term health and development.

When feeding, the social hierarchy, or pecking order, will develop within a group. The largest or dominant lizard will feed first, the others only when it has finished. Similar behaviour is seen with basking spots. If the dominant animal moves to a basking area, any underling will move away to avoid confrontation.

**BREEDING**

A Bearded Dragon becomes sexually mature between its first and second year. Both sexes have spines on the head, neck and throat, but only the males develop the coal-black throat and tail tip during courtship, or when involved in ritualistic fighting over territory. This will be accompanied by much 'head bobbing'.

A healthy female may lay six clutches of eggs in a season, averaging 20 eggs per clutch. These are buried deep into the soil – up to 20cm (8in) – or into a laying box containing vermiculite. They need to be incubated at 29°C (84°F) for about 60 days. After producing eggs or live young, herptiles benefit from a period of isolation and a nutritional diet in order to recover adequately for future breeding.

**▲ Regular handling of juvenile Beardies should ensure that they remain tame throughout their life.**

# BLUE-TONGUED SKINK

*Tiliqua scincoides*

**True ground dwellers, these snake-like lizards like to scavenge amongst the tropical leaf litter, brush land and suburban gardens of their native habitats of north and east Australia and Tasmania.**

At least 1000 species of skink have been discovered so far. Most of them are insectivores. They inhabit a wide range of habitats, but are more commonly found in the tropics. In additon to the Blue-tongued Skink there are a variety of smaller skink species available to the herpetologist. However, only a few are widely available from captive bred sources. Many may be kept communally alongside other small lizards. Some skinks are native to North America and others are found in Mediterranean Europe. Generally, all skinks have polished scales, flattened and elongated bodies and possess reduced limbs.

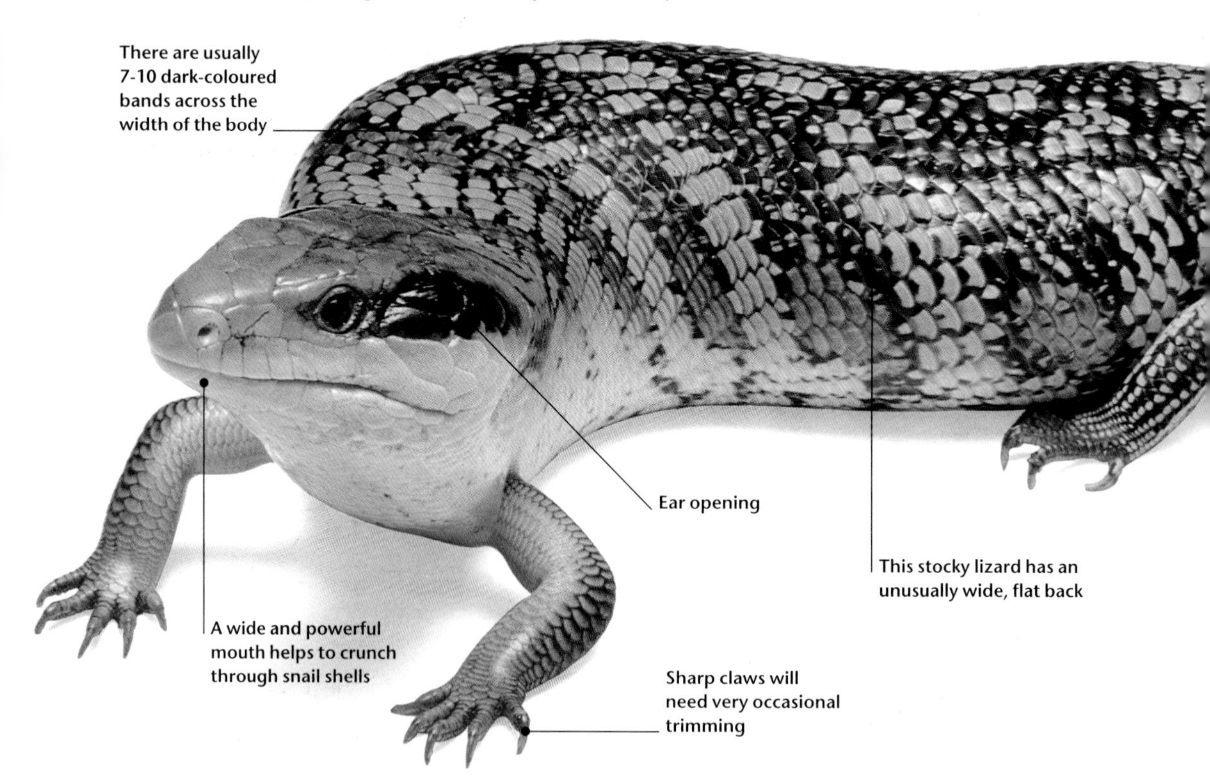

Blue-tongued skinks, as their name suggests, possess a large blue tongue. The tongue can be flattened and rippled inside an open mouth which, when accompanied by a hiss, acts as a very good deterrent to would-be predators. Although totally defenceless, a bite from an untamed lizard would certainly cause some pain. However, regular handling of your skink will ensure it never feels threatened in captivity and the tongue will be rolled in and out of the mouth as it is stroked on the head or under the chin. Its close relation, the Pink-tongued Skink – which has a bright pink tongue – is not as readily available in the UK as the blue-tongued species.

## Pink-tongued Skink

*Tiliqua gerrardi*

The Pink-tongued Skink is also an Australian species and will grow to 45cm (18in). Unlike many lizards, the tail of this species is prehensile. Active mainly at night, they enjoy climbing in the lower branches of shrubs in search of snails, slugs and ripe fruit. They should be cared for in the same way as the blue-tongued variety. Males and females should be housed separately to avoid fighting and only introduced for breeding. They become mature at betwen 2-3 years of age. This species is only sometimes available, and rarely from captive-bred stock.

At birth the live young can be quite large, measuring approximately 15cm (6in) in length. Once mature, a fully-grown adult can reach a length of 50cm (20in). A young Blue-tongued Skink is best purchased at a few months of age, when it is approximately 15-18cm (6-7in) long. In captivity, these skinks can live for up to 10 years.

**◀ Highly polished scales, a pointed head and small legs are all characteristics of the family *Scincidae*, of which Blue-tongues are one of the larger species.**

Males usually have a broader-based tail to encompass the twinned reproductive organs that are inside

## CAUTION

**Ingested substrate can cause terrible problems for all herptiles. If ingested in sufficient quantities, it can cause serious damage to the stomach and digestive tract. Feed from a dish or bowl where possible to prevent particles sticking to food.**

▲ **This spacious vivarium provides everything a growing Blue-tongued Skink needs.**

**CREATING THE RIGHT ENVIRONMENT**

A dry, woodland-type of habitat with plenty of hiding places and retreats suits these bulky lizards. A vivarium for a single skink or adult pair should measure at least 120 x 30 x 30cm (48 x 12 x 12in). Since they actively forage for food, a delicately planted vivarium will be quickly destroyed, so I would suggest you use logs of cork bark and sturdy lumps of wood or rock, securely positioned so that they cannot overturn and injure your pet. Wood chips, peat and leaf litter will help to create an attractive display within the vivarium.

A maximum daytime temperature range of 25-35°C (77-95°F) would be preferred, with a drop at night to 18-20°C (65-68°F). The combination of a heater pad at one end and a thermostatically-controlled spotlight, wired-up to a timer switch, will warm the tank and create a daytime basking area. The light should be timed to switch off at night, thereby automatically reducing the temperature. A daily spraying is much appreciated and so is a shallow, sturdy bowl of water sunk into the substrate.

Blue-tongues can be housed with other herptiles of a similar size. My own adult skink, Sydney, lives with another Australian lizard species, a Bearded Dragon called Barry.

**HANDLING**

These lizards thrive in captivity and are usually easy to tame. They can be handled a great deal without obvious signs of distress, although individuals do vary in temperament. Sharp claws may need occasional trimming.

**FOODS AND FEEDING**

To encourage activity, food should be scattered about, rather than given in a bowl. My skink 'Sydney' is fed this way, but the vivarium requires regular cleaning to remove uneaten foodstuffs and soiled areas. For these scavenging omnivores, snails are a favourite food. They will be tracked along their trail, grabbed by the shell and smashed against a log or the side of the vivarium, before the snail itself is swallowed and most of the shell is discarded.

**AMAZING FACT**

The Pygmy Blue-tongued Skink, a diminutive relation of *Tiliqua scincoides*, was thought for many years to be extinct. Last seen in 1959 in a shed in an Adelaide suburb, this lizard was recently rediscovered, alive and well, close to Adelaide.This lizard only grows to 16cm (6.5in).

Keeping a Blue-tongued Skink may make you very popular with your gardening neighbours because they will have a handy outlet for all their unwanted pest snails. However, avoid using any snails collected from areas where pesticides may have been used.

Besides snails, these skinks will eat cat or dog food, fruit – they like to lick wet juicy fruits such as mangoes – some vegetables and boiled eggs. My own skink is partial to bananas and dog food mash. Richer, processed foods such as cat or dog food should be used sparingly rather than becoming the main diet to reduce the risk of obesity. Crickets and other live foods may hide or out-run these lizards in the vivarium, but cooling the insects in the fridge and then careful hand feeding is an option. Do be careful, however, because a bite from a skink will hurt and may require first aid.

Offer your pet fresh food 3 times a week – uneaten foods, especially meat, should be removed within 24 hours. Garden snails may be collected and maintained in plastic pet homes for use over the winter period.

**BREEDING**

Although the sexes are not easily distinguished, male skinks usually have a wider head and also a broader-based tail to encompass the twinned reproductive organs that are inside. Mating is initiated by the male biting the female behind the head on the neck area and twisting his body for cloacal contact.

**▶ Although more expensive than other herptiles, these large and impressive lizards, with their bright blue tongues, are easy to maintain.**

If successful, the female will give birth to between 6-25 fully-formed young some 4 months later. The exact litter size will depend very much on the size, experience and health of the mother.

Already large at birth, the babies can be fed a diet similar to that of their parents, but the food should be finely chopped and smaller live foods should be offered. The young may be housed together initially, but they should be separated once fighting and bullying start.

Researching this book has produced some interesting findings. Having worked with, and kept reptiles for many years, I was very surprised to read that breeding records for this species go back at least 100 years, when gas lamps and paraffin heaters were used to heat vivaria. The Reverend Bateman in his book, *The Vivarium*, published in 1897, states that London Zoo and private keepers had breeding success with blue-tongues and that they cost 'about 30 shillings when in market'.

# VEILED CHAMELEON

*Chamaeleo calyptratus*

**The chameleon's amazing ability to change colour is much misunderstood. Emotions and territorial posturing, as well as camouflage, are responsible for the rainbow-coloured displays.**

Chameleons are unlike any other lizard in habitat and appearance. With their unique independently moving eyes and projectile tongue, they will always be a favourite with herpetologists. A few species are now beginning to be bred in captivity for the pet market. Imported specimens rarely survive long, mainly due to the stress of travel and delays at airports and customs. The Veiled Chameleon is now captive bred in large quantities and would appear to be one of the most adaptable species, ideal for the pet keeper willing to invest time and effort.

The Veiled Chameleon, sometimes called the Yemen Chameleon, is native to Saudi Arabia and Yemen. Adult males are usually larger than females and can measure up to 60cm (24in), with tail extended. Females usually grow to a maximum of 33cm (13in).

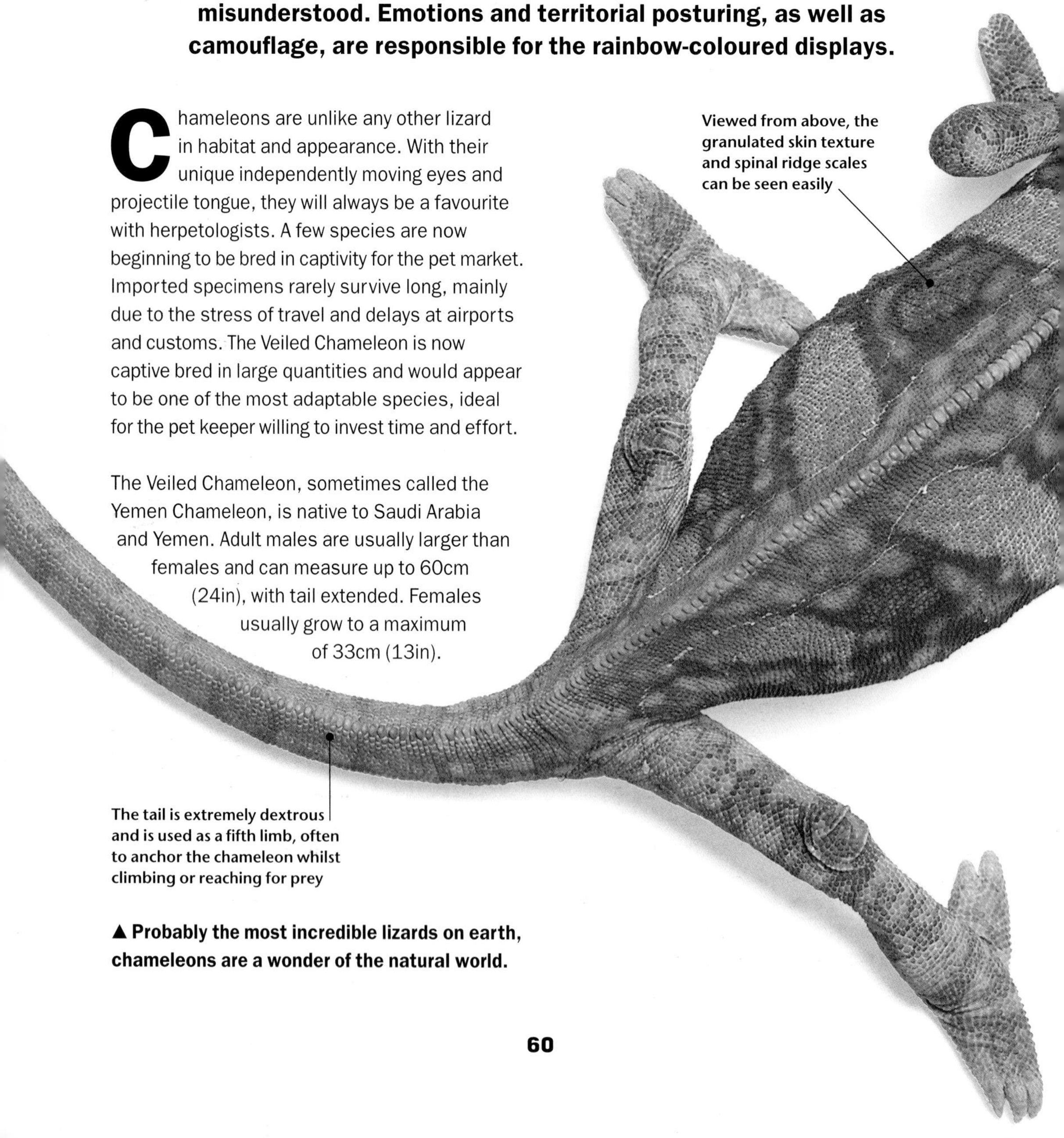

Viewed from above, the granulated skin texture and spinal ridge scales can be seen easily

The tail is extremely dextrous and is used as a fifth limb, often to anchor the chameleon whilst climbing or reaching for prey

▲ **Probably the most incredible lizards on earth, chameleons are a wonder of the natural world.**

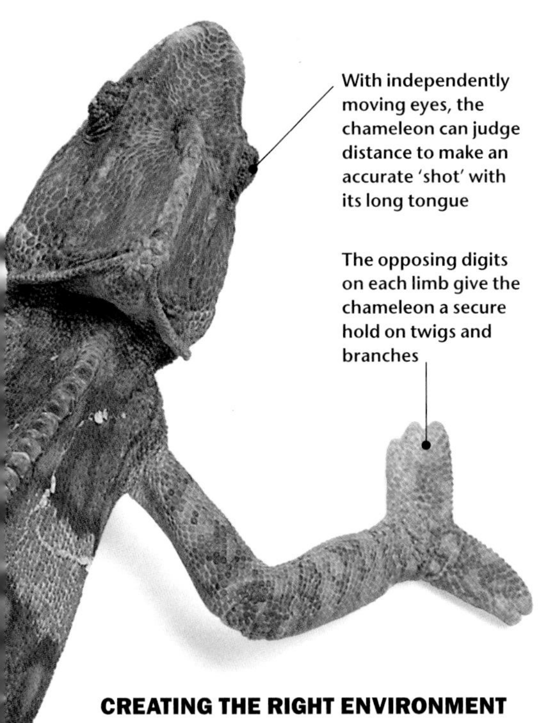


## CREATING THE RIGHT ENVIRONMENT

Larger vivaria suit chameleons best and I would recommend a minimum size of 92 x 92 x 51cm (36 x 36 x 20in) to house a single specimen. Veiled Chameleons are very aggressive towards one another, and adults should be housed individually. As bush or shrub dwellers, it is preferable to offer your chameleon living plants or at least some plastic plants and branches on which it may climb.

One or more spotlights, to create a basking area, will ensure that your pet can warm up properly in the morning before spending the day looking for prey. A temperature of 30°C (86°F) in the basking area, down to 20°C (68°F) in the cooler, shaded part of the vivarium should be available to allow your lizard to thermoregulate with ease. At night the temperature can drop safely to 15°C (60°F). A fluorescent light tube with high UVA/UVB content will ensure good plant and reptile growth.

Chameleons need to drink and eat regularly. Water is preferred in droplet form; indeed, chameleons tend to prefer habitats where heavy dew, or condensation, occurs naturally, since they drink quite a lot in comparison to other reptiles. Mist spray daily and/or provide a pool with water circulating and dripping off branches back into the pool. Good ventilation is essential to the successful, long-term care of these chameleons. A small fan helps to circulate and freshen the air, but it must be placed out of reach of the vivarium inhabitants.

## HANDLING

Many Veiled Chameleons tolerate handling relatively well, especially if it involves hand feeding. Start handling your pet while it is still quite young to condition it to this sort of contact on a regular basis. Never pick up the chameleon, as this will cause your pet great offence. Instead, allow it to climb up onto your hand and fingers. If permitted, it will continue to climb until it reaches your shoulders and head.

## AMAZING FACT

Chameleons are famous for their colour-changing ability. They possess pigment-filled cells called chromatophores just beneath the skin's surface. These rapidly expand or contract causing the colours on the skin surface to change. This behaviour is triggered by external stimuli, primarily the need for concealment or camouflage. Colour fluctuation can be influenced by the lizard's emotional state – angry chameleons will go black. Colour changes may also indicate a territorial display, willingness to mate, state of gravidity, and subjugation or low status in a community.

**FOODS AND FEEDING**

Chameleons enjoy feeding and their method of stalking prey to within range of their long, extendible tongue is fascinating to observe. All live foods can be tried, but flying or climbing species are preferred by these arboreal animals. Variety is the key to provide the range of vitamins and minerals required for healthy development. Some garden-caught food is advised to boost most herptile diets. Larger specimens may accept pink mice, a highly nutritious option.

One method of supplying live foods without the insects dispersing into the vivarium is to attach an opaque pet home, or similar steep and smooth-sided bowl, to a feeding branch. This will position food within reach of the chameleon and its tongue, without enabling the bugs to reach the branch to escape. Make sure that the bug home is well stocked and that the bugs are well fed – the chameleon will use this 'service station' for a top-up. Try hand feeding by placing a bug in between your extended forefinger and thumb.

## Panther Chameleon

*Chamaeleo pardalis*

Native to Madagascar, Panther Chameleons prefer warm and humid conditions and enjoy climbing on small bushes and shrubs. It is one of the most colourful species of reptile, with males showing the greatest colour range – anything from turquoise to fire red and lemon yellow. They are also quite large, with females growing to 33cm (13in) and males up to 51cm (20in).

Second only to the Veiled Chameleon, these creatures are ideal to keep as vivarium pets. A temperature range of 26.5-35°C (80-95°F) should be supplied and full-spectrum fluorescent lighting is essential for the healthy development of both the chameleon and any housed plants. House males separately because they are very territorial and will fight and injure each other. If a large vivarium is available, you may keep breeding trios together – two females to every male. Unlike the Veiled species, Panthers are insectivorous, eating almost any of the commonly available live foods and the odd pinky or small lizard (vertebrate food should not be fed live to your pet).

Unusually, the Veiled Chameleon is an omnivorous creature and will accept a wide range of leaves, fruit and vegetables – these should be dusted with vitamin and mineral supplements.

**BREEDING**

Veiled Chameleons are sexually dimorphic, that is, the sexes differ in appearance. Males have a larger casque (a helmet-like structure on their heads), small 'spurs' on their back legs and are generally bigger than females.

They are prolific breeders and egg layers, and a female may lay clutches numbering between 25-80 eggs, 3-4 times a year. The clutch size, as with most animals, is dependent on the size, age, experience and health of the mature female. This productivity does take its toll eventually and females rarely survive much beyond their fifth clutch. The eggs are buried in soil or sand and should be placed in vermiculite and incubated at 27-29.5°C (80-85°F) during the day, with a small drop at night to 24°C (75°F).

Other aspects of chameleon husbandry are too complex to go into at length, but put very simply Veiled Chameleons should be kept separately and only introduced for mating. Only juveniles should be housed together, and then only for a few months, since they reach sexual maturity between 4-6 months of age. Young females can be mated from about 5 months of age.

Juvenile chameleons, like most herptiles, grow rapidly and it is in these first crucial months that the most varied and well-supplemented diet will enable them to grow up into healthy, mature specimens. They need a larger range and variety of insect prey than the adults which tend to enjoy a more omnivorous diet.

Drosophila and other winged prey are particularly relished by juveniles, who can devour very large quantities at this tender age. Wild caught insects help to ensure a beneficial supply of nutrients.

**▲ Of all the species of chameleon available in captivity, the Veiled Chameleon is taking the herpetological world by storm. It is a breedable, attractive and interesting species to keep, which thrives as a pet under the right conditions.**

# UROMASTYX

*Uromastyx acanthinurus*

**Also known as Dabb Lizards, these unusual looking creatures are found in the southern Mediterranean and in the Middle East. They spend much of their time scrambling over rock piles and dunes.**

Growing to 45cm (18in) in length, this is a large, heavy-bodied lizard with a head not unlike a tortoise. Because of its size, I would recommend that only a novice with enough time and space should keep this species. Their colour develops as juveniles and can be anything from yellowish to red in tone with a dark brown base colouration. These lizards tend to live for 7-10 years, but 15 years is a possibility.

### CREATING THE RIGHT ENVIRONMENT

For a single adult a vivarium measuring 120 x 50 x 50cm (48 x 20 x 20in) is ideal, furnished with rocks, play sand and desert-style decor.
Full spectrum fluorescent lights and a safely-positioned, and protected, ceramic heater are both necessary to ensure full maturing and successful breeding of these lizards. The hot end of the vivarium should be heated with a powerful 60 watt or 100 watt infra-red lamp controlled by a thermostat so your pet can bask at temperatures around 35°C (95°F). Another part of the vivarium must be considerably cooler, 20-22°C (68-72°F), so they can thermoregulate as required.

Their native desert environment is very hot and to escape excess heat they will retreat into cool burrows. Drainpipes or artificial rock hides are suitable for providing retreats. Cork bark is ideal because it is light and can be pegged or siliconed into a range of structures. Ensure your rocks or shelters cannot collapse and injure your pet.

Your Uromastyx should be mist-sprayed occasionally. Once every four weeks give them a bath in clean, tepid, shallow water to allow them to re-hydrate if required, and to aid the periodic process of sloughing.

### AMAZING FACT

These lizards wedge themselves into crevices to make it very difficult for a predator to extract them. When they are too cold and without the energy to flee they are likely to use their club-like tails as a weapon. The vigorous blows can wound any would-be predator, even venomous Horned Vipers.

### HANDLING

A mature, wild-caught specimen is capable of giving a bad bite but, in general, this species seems to be of a pleasant character. If acquired as a captive-bred juvenile, they will tame easily and quickly. Encourage your pet to come to you for food. Move slowly and deliberately so that it is not startled into running away. Eventually it should sit on your hand – at least for a while! Like other medium-sized lizards, you can pick them up gently with one or two hands.

## FOODS AND FEEDING

Virtually herbivorous, their diet should consist of mainly green leaf foods with pieces of vegetable and fruit added. Carrots, peas, beans and hibiscus leaves are also enjoyed. Some insect foods may be accepted. Feed every other day and remove uneaten foods on the second day to minimize fouling. Supply food on shallow trays away from main basking spot to avoid desiccation of the food.

This species is likely to aestivate, at least for short periods. It would appear that this correlates to its behaviour in the wild and you should not worry unduly if your Uromastyx rests without eating for 1-2 weeks in one of its favourite retreats.

## BREEDING

Trying to breed Uromastyx is difficult but not impossible. Mating occurs in the spring and eggs are laid in May or June. Eggs should be incubated in a dryish medium of vermiculite or purlite at 30-33°C (86-89°F) for approximately 12 weeks. When breeding any herptile, try to ensure the adults are given a very nutritious and calcium-rich diet to prepare them for the physically exhausting process of producing and laying eggs.

Uromastyx are quite aggressive to each other and are best kept singularly or in trios – one male to two females – but only if very large accommodation is available. After mating, the female should be housed on her own with a suitable amount of substrate to encourage egg laying.

**▶ Only the healthiest specimens should be considered for a pet. Avoid any thin, inactive individuals and watch out for damaged or swollen digits, a thin bony tail and a dirty vent.**

# GREEN IGUANA

*Iguana iguana*

**These 'tree chickens', as the West Indians call them, spend their time grazing on leaves, fruits and flowers in lush tropical vegetation. Good swimmers, they will dive into water to avoid predators.**

Native to Central and South America, the Green Iguana is probably the most commonly kept large lizard but it is only suitable for those pet keepers with considerable space and time available to them. Not only are they large, they can live for 10-15 years in captivity. Although basically green, Green Iguanas can display a range of colours. My iguana (pictured right) has a remarkably blue head and a colourful dewlap, or throat pouch.

When buying an iguana it is important to get one as young as possible because early handling will help to foster a good relationship between you. Look for a lizard that is not too bony at the tail base, is alert and not sluggish, and preferably one that looks interested or will come towards you when food is offered. Distressed or scared individuals will shoot off in flight at the first sign of trouble.The temperament of lizards vary and you must hope to choose one that will become used to, and enjoy, your company rather than attempt to bite or flee.

**CREATING THE RIGHT ENVIRONMENT**

Iguanas need more space as they grow, with height being the most important factor to consider, although they will use the ground area too. Start off with, at the very minimum, a 92 x 38 x 30cm (36 x 18 x 12in) vivarium for your baby iguana. Adult iguanas require a custom-made vivarium measuring a minimum 120 x 180 x 120cm (48 x 72 x 48in).

▲ **As it grows, this hatchling iguana will require considerable space to allow for its healthy development.**

Provide sturdy branches, plastic plants for shade and a large, non-spillable water dish. Cork logs make ideal retreats and your iguana will probably sleep in one, especially if it is nice and warm. A good hot basking spot, 30°C (86°F) is required for this diurnal reptile. A full spectrum tube is also essential for maintaining your lizard's longterm health.

▼ These truly magnificent creatures are suitable only for those pet keepers with the considerable space and time to tame and care for them adequately.

▲ **The long sharp claws and the elongated toes enable iguanas to climb into the rainforest canopy.**

## HANDLING

Iguanas are not easy to handle unless they are very tame. Claws grow long and lacerate human flesh so you should trim them occasionally, taking care to avoid the 'quick', or living part of the claw. Very tame Iguanas may be walked on a lead or perched on a shoulder, but most never tame to this extent.

Less tame individuals need professional restraining – using long leather gauntlets to protect yourself is recommended. The iguana should be held firmly behind the neck and above the rear legs and the tail should be tucked under your arm to avoid its painful whip. Juveniles are likely to drop their tails for defensive reasons but this ability is lost once iguanas mature.

## FOODS AND FEEDING

Adult iguanas are herbivorous and will eat a wide variety, and quantity, of foods. It is important not to overfeed your pet on richer, 'preferred' foods since, in the wild, the bulk of its diet is made up of poor quality foods. Adult Iguanas should be fed every other day or so. A large stainless steel bowl filled with a mixture of fruits, leaves and vegetables is sufficient. Depending on the season, you could include the following: chopped

**AMAZING FACT**

In the Caribbean Iguanas are called tree chickens, and an adult, once cooked, can feed a family of four.

lettuce, corn, cucumber, flowers, dandelion, cooked potato, apple, pear, kiwi fruit, banana (including skin), grapes, green beans, kale, broccoli, melon, plums and mango. Remove the food bowl on the second day and replace with more food, at the latest, on the third day.

Juvenile iguanas will eat much the same thing, but in a smaller, finely-chopped form and adding some live foods – crickets, locust hoppers and waxworms – is necessary. Taming and feeding your juvenile goes hand in hand. You really must invest considerable time in taming your iguana so that you can both enjoy a good relationship.

**BREEDING**

During copulation, a male Iguana will firmly bite and hold the female at the neck or head. By twisting his body partially under hers, mating is possible and can last from a minute to nearly half an hour. A gestation period of 50-90 days is typical, and the female will dig a long chamber about 50cm (20in) deep and 1-2m (39.5-79in) in length. Up to 40 eggs are deposited over several hours and, after burying them, they are left to incubate in the warm soil. Under optimum conditions they hatch about 10-15 weeks later and the youngsters spend the first months of life chasing invertebrates and feeding on leaves and flowers.

Iguanas are farmed in Central America and bred in zoos and institutions in the UK and USA. Captive breeding is only likely in very large enclosures. It is unlikely for most pets and, indeed, as they grow too large for their vivaria, many await re-homing or end up in the local zoo or wildlife park. Home breeding is not recommended.

**CAUTION**

**Many pet stores sell the tiny, emerald green juveniles that only measure 25cm (10in). Bear in mind, however, that they will eventually grow into a formidable 200cm (79in) long adult with a strong tail that will lash at anything or anyone threatening it.**

**◀ The enlarged scale, clearly visible beneath the ear opening, coupled with jagged spines along the length of its back and on the edge of its extendable dewlap, can make an Iguana look quite formidable.**

# SPUR-THIGHED TORTOISE

*Testudo graeca*

**These hardy creatures are native to the Mediterranean and may be found in Spain, Greece and Turkey. They are quite well suited to the British climate but do require a little help to grow, thrive and breed.**

Thousands of Spur-thighed Tortoises are bred every year in captivity, both in the UK and USA. A captive-bred baby measuring 7.5cm (3in) is an ideal acquisition. Initially, it will need to be kept in a vivarium indoors in winter, and kept in an outside enclosure during the summer.The shell colour can vary from yellow to black and they reach 20cm in shell length.

## CREATING THE RIGHT ENVIRONMENT

For all but the youngest juveniles, who need to be penned, tortoises may be kept in a walled or fenced garden for much of the spring and summer months when the garden is frost free. A hutch provides shelter from rain and excesses of temperatures. It should be fitted with a heater pad or infra-red lamp to keep chills at bay and to enable

**▲ A youthful 15 year old, this female tortoise is of prime breeding age. You should expect at least 30-50 years from a captive-bred animal.**

## HIBERNATION

Tortoises should be hibernated during early to late autumn. Many people have found that their tortoises hibernate naturally outdoors, where they dig themselves into a sheltered spot, often under a tree. Depending on their choice of resting place, and the winter weather, they may do well but are exposed to all sorts of risks. These include severe weather, damp conditions and being dug up by predators. It is much better to hibernate your tortoise in a cold (unheated) but frost-free cellar or garden shed.

Tortoises must empty their digestive tract before hibernating. A gradual reduction in temperature a few weeks prior to hibernation should work because the tortoise's appetite declines but it still digests food in the system. An underweight tortoise should not be hibernated and if in doubt, visit your vet or specialist pet centre for advice. I hibernate mine in a hutch furnished with straw (avoid hay because cut grass can cause lung problems) from around September/October to March, depending on the severity of the seasons. Hibernated tortoises should be checked on a regular basis. Upon waking, a warm bath helps them to rehydrate and clears the eyes, nostrils and mouth of any debris and usually stimulates a bowel movement prior to a feeding. It is not advisable to hibernate baby tortoises until they have a plastron length of around 13cm (5in).

the tortoise to be active on cooler days. (Make sure the external fitting is safely installed by a professional). Locate the hutch in a sheltered spot, away from direct wind and sunlight. Ensure that ponds and other deep water features are fenced off to prevent your tortoise from drowning.

Despite their reputation, a healthy tortoise is a fast creature, which will crash through flowerbeds to smell plants before trying them. They can be deterred from doing so by placing 25cm (10in) high boards around the garden. In the morning, they like to orientate themselves towards the sun and a south-facing slope is appreciated to help catch those all-important rays. They soon learn the location of their hutches, the best sunbathing spots and favourite places to rest, feed or sleep.

**▶ This two-year-old Californian-bred Spur-thighed Tortoise is being held carefully using two hands.**

### HANDLING

Tortoises tame quickly, especially when offered tasty titbits. The tendency for the head to be withdrawn gradually decreases as they become used to human company. Tortoises can be held firmly from either side, with thumbs on top of the shell and fingers underneath the body – this is the usual method.

## FOODS AND FEEDING

The Spur-thighed Tortoise is a herbivore and will be satisfied entirely with most fruit and vegetable matter you offer. Vitamin supplements should be added to the food to promote healthy shell formation, although access to graze in a garden and exposure to sunlight will usually fulfil most vitamin and mineral requirements.

My own tortoises are offered chopped fruit and vegetables on roofing slates (flat and easy to clean) placed in the garden. Consider offering the following, making sure you provide a good selection: Brussels sprouts, chickweed, clover, runner beans, mangetout, grated carrot, cress, peas, broccoli, bindweed, lamb's lettuce, cauliflower, rose petals, cucumber, and soft fruits (not citrus). They can also forage – I positively encourage dandelions, a favourite tortoise food, but not every gardener's favourite weed.

Food preferences will vary, but experience will help to guide you. Lettuce should only be a small percentage of your tortoise's diet, no matter how much your pet enjoys it, since it is of low nutritional value. Although normally herbivorous, some tortoises seem to enjoy a small amount of meat in their diet. In the wild tortoises will eat available carrion, such as dead birds and snakes. Be careful not to overfeed on meat and restrict it to 1-5% of their total diet.

Since tortoises come from the world's naturally arid environments, the majority of the moisture they require actually comes from the foods they

**▼ Providing a variety of foods helps to ensure that your pet tortoise will grow and develop into a healthy adult.**

consume. However, particularly during very hot, dry periods, additional water may be needed and tepid water may be dribbled or poured over your tortoise's head. They may then take the opportunity to drink a small amount. Alternatively, you may offer your pet tortoise some water in a shallow dish.

**▲ This large 'pet home' unit is ideal for rearing baby tortoises, keeping them warm and well fed.**

**BREEDING**

Adult male and female tortoises are easily distinguished. The female has a shorter tail and a flat underside, while the male has a longer tail and a concave underside which enables him to mount the female for mating.

The best male/female ratio for breeding tortoises is one male to three or four females. Males dominate their females. By ramming his shell against hers, biting her legs or tail, the male will force her to give up trying to escape his advances and enable mating to take place. Mating is unmistakable, with the sound of much clanking and bashing of shells prior to the male vocalizing once mating is underway. The rough courtship of the male can cause considerable distress and some injury to the female, whose legs may become sore and bloody as a result of his continual advances – hence the recommendation for a harem of three or four females to divide the male's attention amongst them.

Eggs are buried about 15cm (6in) deep in the soil and so well covered by the female that, unless egg laying is actually witnessed, it may be impossible to locate the site. Eggs must be removed, without turning, and placed in vermiculite at between 28-32°C (82-90°F) at a relative humidity of 70%. Hopefully, these will then hatch approximately 75-85 days later.

For their first year of life, keep baby tortoises in an indoor vivarium under full-spectrum light tubes at 25-28°C (77-82°F). Move them to small, contained runs in the garden during the warmest months, where access to real ultraviolet rays will help with healthy shell formation. At this stage they are still vulnerable, so be aware of possible predators.

**AMAZING FACT**

Tortoises helped Darwin to develop the revolutionary idea of evolution. He noticed that, although very similar in appearance to those on the South American mainland, the tortoises on the Galapagos Islands were much larger. He worked out that over millions of years small changes take place in each new generation due to the different environmental conditions faced by a species. These offshore tortoises had gradually developed into Giants.

# Reptiles 2

## Snakes

These predators belong to the largest order of reptiles – the Squamata. Secretive by nature and often solitary, they spend much of their lives concealed in hollow logs or mammal burrows. Some of the more commonly kept species include Milk Snakes, Royal Pythons and Garter Snakes.

All snakes are predatory vertebrates. They are found throughout the warmer parts of the world and they thrive in numerous habitats, including open sea, deserts, underground or high up in tree canopies.

Snakes, like lizards, chelonia and amphibians, are ectothermic. This means that they rely on their environment to satisfy their heat requirements. In extreme temperature conditions, many temperate species hibernate in winter and tropical species aestivate in summer. Some species of Garter Snake are only active for three months of the year because their natural environment is extremely cold.

Snakes have evolved from, and can be described as, specially modified lizards which have adapted to a limbless existence. Their scales may be flat, overlapping or keeled with a raised ridge in the centre to allow them to grip surfaces successfully.

Periodic sloughing replaces the old skin. In snakes the old skin is always discarded and is not consumed as it is by most amphibians and many lizards.

The structure of the skeleton varies from species to species. It is more elongated in snakes than in any other reptile and consists of up to 400 separate vertebrae, mostly attached to long curved ribs. The backbone is extremely flexible when moved from side to side, but has little vertical movement.

Snakes move by contracting and expanding the body lengthwise while simultaneously gripping the ground by pressing the body and ribs onto large wide ventral scales. Other methods of movement employed by snakes include side winding and a concertina style of movement. Some primitive snakes still display visible remnants of legs possessed by their lizard ancestors.

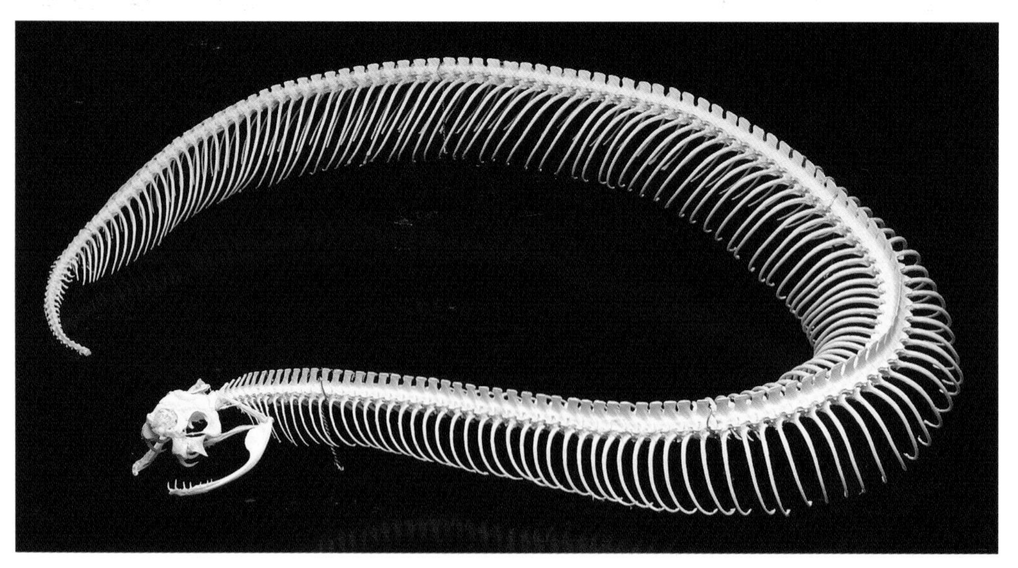

**▲ Typically for reptiles, the bones of snakes do not stop growing once they mature and continue to grow throughout their lives.**

**▶ Snakes are seen by many people as beautiful and fascinating creatures that should be encouraged into the household as pets. Colourful species, such as this Corn Snake, are bred to the extent that they may soon be described as domesticated animals.**

The upper and lower jaws are loosely connected to the skull and may be expanded greatly – most species eat prey much larger than their head/neck size would suggest. All snakes are carnivores and conume food whole. Food is either swallowed alive constricted within coils and asphyxiated, or killed by an injection of lethal venom. Snakes are unable to chew or cut food into portions and prey is simply swallowed in one piece, usually head first.

As predatory reptiles, a snake's success as a hunter is dependant on the combined use of the eyes and tongue. Snakes cannot open or close their eyes; in effect, the eyelid is shut but transparent. A snake's visual perception is very different from our own. We can see individual objects, whereas snakes can only see or perceive an object when it moves or changes in some way. The tongue collects tiny airborne particles which are analyzed by a special structure called the Jacobson's organ in the roof of the mouth. A snake's tongue is forked and it effectively tastes or smells the air in 'stereo', each fork providing some sense of direction.

Snakes have no visible ear opening and do not hear sounds in the way that humans can. They are sensitive to ground vibrations which are felt through the lower jaw.

The tail of a male snake is often longer than that of a female. It is also generally wider at the base where the twinned or hemi-penis is enclosed. Females lay a leathery shelled egg. Some species, mainly those from temperate areas, incubate their eggs internally and give birth to fully formed young. Snakes rarely provide parental care – the majority of eggs are simply laid, buried and then left. A few species, however, such as some cobras and pythons, do guard and protect their nests.

**▶ As reptiles grow, they will periodically slough their skin. This process normally starts around the head. A juvenile snake may shed its skin every month, but an adult will do so only 3-4 times a year.**

# KINGSNAKES

*Lampropeltis getulus*

**Kingsnakes are an American species found in a large area ranging from Canada in the north to Ecuador in the south. They are very popular pets due to their size, great colour and variable patterns.**

All Kingsnakes are ophiophagous. Normally they will feed on other snakes in the wild, but are generalized carnivores in captivity. They feed well and will eat almost any prey – birds, lizards, frogs and rodents. They are so popular that many thousands are bred by hobbyists and professional herpetologists every year for the pet trade.

The common subspecies grow to about 120cm (48in) and live for 10-15 years. Subspecies to look out for include the following:

| | |
|---|---|
| **Californian Kingsnake** | *Lampropeltis getulus californiae* |
| **Florida Kingsnake** | *Lampropeltis getulus floridana* |
| **Mexican Black Kingsnake** | *Lampropeltis getulus nigritus* |

**CREATING THE RIGHT ENVIRONMENT**

House Kingsnakes separately because they are cannibalistic. A clean, dry vivarium heated to 26.5-32°C (80-90°F) at the warmest end and 24°C (75°F) at the cooler end would be ideal. These snakes rarely climb so the height of your vivarium is far less important than the provision of hides in both the warm and cool parts of the unit. A vivarium measuring 90 x 30 x 30cm (36 x 12 x 12in) is more than adequate. There are no supplementary lighting requirements, since background daylight is sufficient.

Paper or wood chips are the most commonly used substrates. The use of rocks, cork bark and a panoramic background can help to create an attractive display. These snakes do climb to a small extent, but branches are not essential. Fresh water should be available in a shallow water dish, large enough for the occasional bath.

▲ **The temperature is controlled by a thermostat and the artificial plants provide a cool retreat that can be regularly mist sprayed.**

## HANDLING

The very youngest Kingsnakes will vibrate their tails angrily, strike at you, attempt to wriggle away, or defecate on you when handled because they think you are a hungry predator. The best course of action is to continue to handle your juvenile a few times a week. Eventually, it will realize that you are not a threat, but a friend offering occasional food rewards.

Most Kingsnakes tame very well and rarely offend their keeper by biting. Most pet keepers, however, will be bitten by accident sooner or later and, apart from the initial shock, your snake is unlikely to cause much damage.

A Kingsnake will happily coil around your hands and wrists as it explores. Remember not to handle your snake after feeding or before a slough.

**▶ Kingsnakes are a highly variable species – some are striped and the darker markings can range from jet black to a chocolate colour. The white markings are also variable, ranging from a creamy yellow to pure white.**

## FOODS AND FEEDING

Kingsnakes will happily eat each other and any other herptile within reach and, therefore, need to be kept singly. Providing you have purchased a healthy captive-bred specimen you should have no feeding problems. They are most likely to refuse food just before a slough and will be at their hungriest just after. For a more nervous specimen leave food in a hide box to make them feel more secure.

A diet of dead rodents is sufficient for all the snake's nutritional needs, assuming it is eating healthy well-fed rodents. Even juvenile snakes will usually accept pink mice. In the USA, many keepers and breeders will also provide their snakes with small lizards, such as *Anolis sp.* and *Sceloporus*, to encourage feeding and to offer some variety in the diet. The cost of lizards in the UK usually prohibits this. Because mice are easily obtained ready frozen, needing only to be thawed and warmed, they provide the bulk of most pet Kingsnakes' diets. Supply food in a size relative to your pet; the bulge created upon swallowing food should never be abnormally large. Feed your snake as much as it will eat in 15-20mins, every 6-8 days.

### AMAZING FACT

Kingsnakes are noted for their unusual ability to kill and consume venomous snakes such as Coral Snakes and Pit Vipers. Although they show little or no ill effects from the bites of American Pit Vipers, immunity to the venom would seem to be variable within the species.

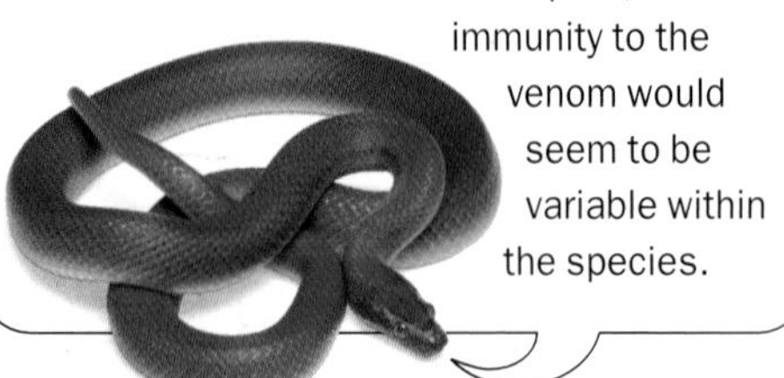

### PREDATOR & PREY

Kingsnakes eat a variety of foods in the wild – other snakes, lizards, birds and even reptile eggs. They are marauding hunters, actively seeking prey above ground and below in mammal burrows and tunnels. They possess the unusual ability to kill more than one prey at a time. The first victim is seized in the jaws, and subsequent prey is pushed against a rock or smothered in coils. In turn, Kingsnakes are preyed upon by raptors and mammals, such as the coyote, as well as other Kingsnakes.

## BREEDING

These snakes are most likely to mate after a period of hibernation. They should be kept for several weeks, without feeding, at temperatures of 12.5-15°C (55-60°F). After this period, raise the temperature to their normal vivarium conditions. When ready for mating, usually after a slough, the female releases hormones to attract the male. Mating snakes should be watched to ensure one does not eat the other.

Return the female to her vivarium. Insert a laying container to provide a secure and comfortable spot for the female to lay her eggs. About 5-7 weeks after mating 5-17 eggs will be laid, each measuring up to 5cm (2.5in) in length and 2cm (0.75in) in width. These should be removed and incubated in vermiculite at 24°C (75°C).

Within 70 days the eggs will hatch into 25cm (9.5in) long babies. Once sloughing has taken place, usually 12-16 days after birth, they will be ready to feed on the smallest of pinkies. At 4-6 months of age they are an ideal purchase as a pet. This species matures at 2.5-3 years.

▲ This chocolate and banana coloured specimen is another variation of the Californian Kingsnake.

▶ The glossy Mexican Black Kingsnake is a very popular choice for the vivarium.

▼ Florida Kingsnakes are one of the heavier, more solidly-built species.

# MILK SNAKES

*Lampropeltis triangulum*

**Noted for their eye-catching bands of colour, they are also the smallest members of the Kingsnake family. Possessing cannibalistic tendencies, they need to be housed individually.**

This American species is found in an area ranging from southern Canada to northern South America. They live in a variety of habitats, ranging from swamps to deserts.

There are about 25 recognized varieties, or subspecies, of Milk Snake and they are the smaller and more colourful members of a group of snakes called Kingsnakes. In Latin their name, *Lampropeltis*, means 'shiny shield'. Milk Snakes are well known for their habits of eating poisonous American snakes and are best kept separately to avoid unfortunate accidents.

Depending on the subspecies, these snakes will grow to 50-120cm (20-48in) in length and will live for approximately 10-15 years.

Milk Snakes are widely available in pet shops and directly from breeders. Look out for those listed below, since they are available from captive-bred stocks and are easy to maintain in captivity.

| | |
|---|---|
| **Pueblan Milk Snake** | *Lampropeltis triangulum campbelli* |
| **Mexican Milk Snake** | *Lampropeltis triangulum anulata* |
| **Honduran Milk Snake** | *Lampropeltis triangulum hondurensis* |

**CREATING THE RIGHT ENVIRONMENT**

Milk Snakes should be housed individually because they have cannibalistic tendencies. They prefer small, secure, accommodation and, since they are unlikely to climb much, your vivarium does not have to be very tall either.

They spend much of their time hiding below ground and, given the choice, your pet is likely to wriggle under the substrate and spend many hours resting. A mixture of wood chips, sand and dried moss is suitable. Add dried pine cones, driftwood, cleaned leaves and cacti skeleton to enhance the appearance of your pet's home.

**▲ This glass vivarium provides a secure and desirable home for a Milk Snake. It has retreats and is warmed from below by a heater pad.**

A 60 x 30 x 30cm (24 x 12 x 12in) vivarium is ideal for a 50-60cm (20-24in) specimen. Establish a temperature gradient, keeping the hottest parts at about 30°C (86°F) and the cooler areas at 21-23°C (70-74°F). Under vivarium heating is desirable to warm the substrate – using heater pads is the most suitable method. Wire them to a thermostat and ensure they cover no more than half of the ground area. No extra lighting is required.

## HANDLING

Most pet Milk Snakes become tame and easy to handle without problems developing. However, some may want to be left alone and they will defecate, writhe about or may regurgitate food if handling is attempted too soon after a feed. Select a relatively tame snake to start with and then patience, persistence and common sense will help you to tame your pet within a short amount of time.

## FOODS AND FEEDING

Always feed these snakes separately to avoid one snake mistaking his fellow house mate for a meal. Never keep adults and young snakes together because the adults are likely to eat their younger friends.

Your recently acquired Milk Snake should be settled into a diet of pinkies and, so long as it feeds well, it will grow quite rapidly. Baby snakes will usually begin to feed after their first slough at 10-14 days. They should then be offered 2-3 small pinkies at a sitting, every 5-6 days. At several months of age (when they are best purchased as pets) they can be offered several pinkies or furries – as much as they will consume in about 10 minutes – every 7-10 days.

Once adult, feed your snake as much as can be eaten in 10-15 minutes, every 6-8 days. The food a snake consumes should leave a visible, but not abnormal-sized, swelling along its body. You many find that after feeding regularly for several weeks your pet may refuse food for a number of weeks. Do not worry, it is simply full and this is quite normal.

## BREEDING

These snakes are not sexually dimorphic, so there is no easy way to sex them. Most snakes will be sexed by the breeder or dealer who is supplying the specimen. For breeding information see the section on Kingsnakes on p. 78.

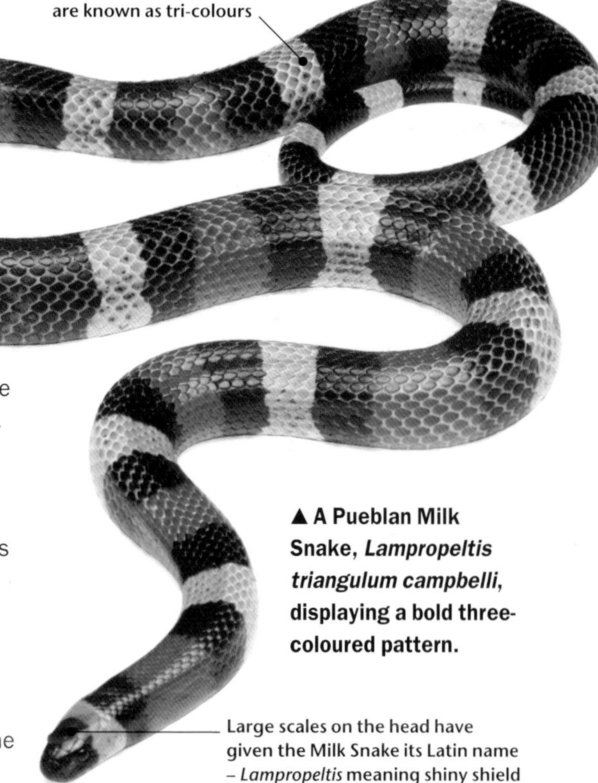

Some Milk and Kingsnakes are known as tri-colours

**▲ A Pueblan Milk Snake, *Lampropeltis triangulum campbelli*, displaying a bold three-coloured pattern.**

Large scales on the head have given the Milk Snake its Latin name – *Lampropeltis* meaning shiny shield

# CORN SNAKE

*Elaphe guttata guttata*

**This colourful American species, also known as the Red Rat Snake, is one of the most commonly kept and bred snakes. They are easy to keep, making them an ideal introduction to snake keeping.**

There is much to recommend Corn Snakes for the first time keeper. They feed well, are usually of good temperament and are very attractive. Now available in a number of colour variations, they have popular names such as Snow and King Corn. Measuring 80-150cm (31.5-60in) once adult, they can live up to 20 years, although 10 to 15 years is more common.

## CREATING THE RIGHT ENVIRONMENT

They require a dry setup with temperatures between 25-30°C (77-86°F) in summer, down to 20°C (66°F) in winter. An adult pair require at least a 90 x 60 x 60cm (36 x 24 x 24in) vivarium. Juveniles should be housed in smaller plastic rearing boxes measuring 24 x 10 x 10cm (9.5 x 4 x 4in) or in a 60x 30 x 30cm (24 x 12 x 12in) vivarium.

**▲ Attractive colourations and good temperament have endeared these snakes to many pet keepers.**

▲ **Corn Snakes require access to fresh water for drinking and thoroughly enjoy the occasional bath. Increased humidity helps snakes slough without difficulty.**

## FOODS AND FEEDING

Corn Snakes feed well and can be given the domestic mouse, *Mus musculus*, for all of their natural life. However, you should be aware that newly hatched Corn Snakes, typically measuring only 23-36cm (9-14in), can have a bit of trouble getting started. Even pink mice can be too large or difficult for them to swallow at this age. In the wild some baby Corns feed on baby frogs or lizards until they grow to a size when rodents can be easily consumed.

A number of substrates are suitable, ranging from plain paper to bark chips and wood chips. Reptile grass is a useful and more aesthetically pleasing alternative to newspaper. A branch for climbing and 1-2 hides must be provided. Make sure a shallow water dish, large enough for the occasional bath, is available. Leaves may be added to make the cage more decorative – freeze these before use to kill potential pests, such as mites or ticks.

If your baby Corn is not feeding, offer mouse legs or a warmed up sliced pinky – not a pleasant job, but satisfying when your snake does start to feed. After a couple of sloughs, your pet should be able to eat pinkies that have been thawed out and then warmed. A snake on this sort of diet is known as a defrost feeder. I would advise buying a Corn Snake when it is a minimum of six months of age. By this stage it should be a good, regular feeder, well past its problem age.

## HANDLING

Baby snakes are usually very defensive to begin with and will hiss and strike out at anything that moves. Once you have one in your hand it should start to calm down; the snake will gradually realize that the scent of a human hand causes them no harm. When picking up an adult snake, support it along the length of the body. You should not dangle a snake by its neck or tail.

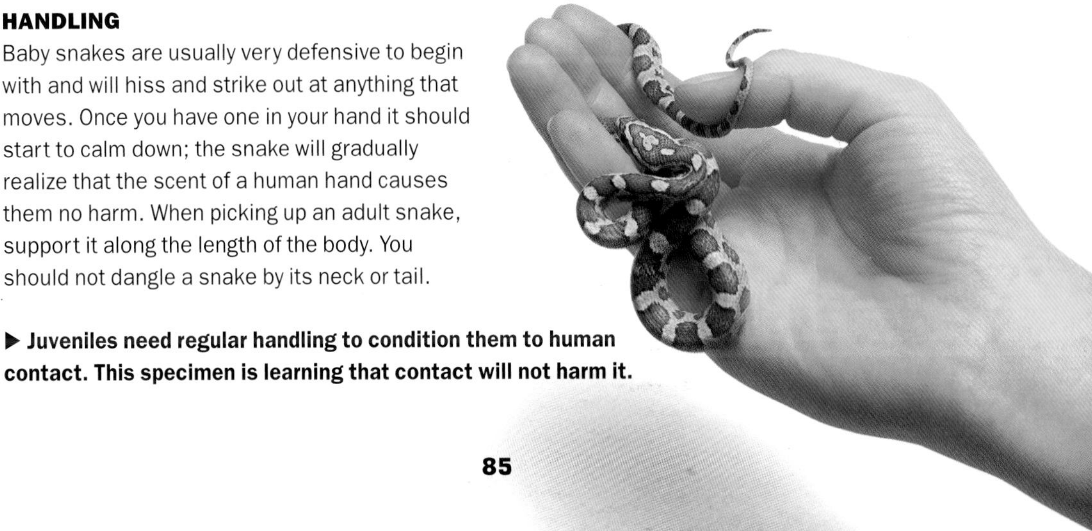

▶ **Juveniles need regular handling to condition them to human contact. This specimen is learning that contact will not harm it.**

Feed hatchlings on very small pinkies every few days. Juveniles will require several pinkies or furries every 7-10 days and adults will require 2-3 adult mice every 2 weeks.

Grown mice should provide all the necessary vitamins and minerals for healthy growth and development. Make sure you only feed domestic mice and avoid using any wild-caught prey. Although your snake will quite happily eat any animal that your cat kills, any animal caught in this way may introduce internal parasites, such as worms.

Corn Snakes kill their prey by constriction. Their usual method is to strike out and bite the prey on the head. The snake then coils around its victim and squeezes, thus asphixiating the prey. For ease of consumption, the prey is almost always swallowed head first. Even when feeding on pre-killed rodents, many pet snakes will still go through the motions of 'killing'. Once they are satisfied that the prey is ready to consume, they will usually smell it from end to end and, by expanding their jaws, they can consume food larger than their head/neck size would otherwise indicate.

**▲ Babies are initially reared in small containers to give them a sense of security. Water, shelter, warmth and food are essential for health.**

**BREEDING**

Most male snakes have longer tails than the females. Measure similarly aged snakes from the vent to the tail tip to ascertain the correct sex of your snake.

Breeding usually takes place during spring and mating is characterized by an intertwined embrace. Mating in snakes provides an ideal opportunity to watch an ancient but effective reproductive process. A keen male will follow a prospective female around the vivarium often biting her and rubbing against the length of her body. Eventually, he will bite her firmly on the neck and pull the weight of his body on top of hers. He will stimulate her by rubbing and making wave-like motions against her cloacal region and mating will take place. Courtship and mating can last for several hours.

Up to 20 eggs are laid in a well-concealed spot in the soil – suitable places are under a dish or a log, or within a hide box. These should be incubated in vermiculite at 28-30°C (82-86°F). The eggs will then hatch after approximately 10 weeks. Corn Snakes mature at 2-3 years, when they are about 80cm (31.5in) long, and can breed until they are about 10 years of age.

**VARIETIES**

The following varieties of Corn Snake are available from captive-bred stocks.

| | |
|---|---|
| **Amelanistic Corn (albino)** | Lacking black pigment in skin and eye, they display a red and orange pattern. |
| **Motley Corn** | Pale orange in colour, they possess an unusual and irregular pattern. |
| **Black Corn** | These snakes display a predominance of black pigmentation and are often completely black. |

Colour variations found in some Corn Snakes are the direct result of the inbreeding of naturally occurring, but rare, genetic mutations. However, inbreeding does carry the risk of lower fertility and other health problems. A normal Corn Snake is, I think, the nicest and best snake to keep, but many people prefer snakes that display unusual skin colourations.

In the wild, many of these unusually coloured snakes would be marked out as easy prey and, as a result, would be much less likely to survive for very long. They are only abundant under captive conditions where such unusual markings are deemed desirable. Within any population of animals such mutations will appear occasionally.

▲ **This juvenile Amelanistic Corn Snake has no black pigment in the skin and the eyes are usually pink or red. They are more sensitive to light and are generally of a more delicate disposition than those with normal colouration.**

▶ **This young Black Corn has black saddles instead of the natural red colouration. Breeders now produce a range of colour variations.**

# HOUSE SNAKE

*Lamprophis fuliginosus*

**These chocolate brown snakes are a common African species and are frequently found in, under or near houses in villages and towns. In the wild, they feed on rodents and lizards.**

This is a relatively new species to be bred and available in the herpetological trade. Although not as commonly available as the more colourful King and Corn Snakes, they are, however, easy to keep and breed. This snake is highly recommended for the novice snake keeper because they are rated as one of the easiest to handle. House Snakes can grow up to 112cm (44in) and will live for 10-15 years.

**▼ At 30cm (12in), this juvenile is ready to go to a new home. Eager to feed and easy to breed it is an ideal first choice for the novice keeper.**

Smooth scales may feel moisturized but never slimy

The milk chocolate colour of body and head contrasts strongly with the creamy belly scales

## CREATING THE RIGHT ENVIRONMENT

These snakes are best kept in a dry, woodland-style setup. The floor should be furnished with a substrate, such as bark chips. Logs, cork bark or other suitable hiding places also need to be provided. They love their hiding places; six of my young house snakes, each measuring 30cm (12in), share a 60 x 20 x 20cm (24 x 8 x 8in) vivarium and when they are all coiled in the hide, their heads often pop out of the opening to survey the scene.

A pair of adults should be housed in an escape-proof vivarium measuring at least 90 x 45 x 38cm (36 x 18 x 15in). Make sure water is available at all times. Insert a humidity chamber. This is a box with a small opening that is filled with moistened moss or vermiculite – it has a high humidity in comparison to the rest of the environment. Alternatively, introduce a section of well-sprayed plastic plants to provide a suitably humid region. The temperature range should be maintained at 25-28°C (77-82°F).

## HANDLING

One of the reasons I recommend purpose-built vivaria, instead of modified fish tanks, is that the best way to approach a snake is from the side

### VIVARIUM SECURITY

A vivarium lock should be fitted to keep your pet in and unwanted guests out. A rubber wedge is also useful to stop the doors sliding open if the unit is not locked.

Glass tanks and vivaria should be placed on a sheet of polystyrene to insulate the glass or wood, and prevent the base from cracking or scratching furniture.

A circuit breaker should be fitted to any electrical equipment – ask in-store or check with a qualified electrician.

**▲ In this artificial snake hide wait two young House Snakes. Much of their time is spent concealed in this way.**

rather than from above. Many predators attack from above and the snake may not have time to 'smell' or 'see' your hand before striking out at this predatory movement. Front opening vivaria will eliminate this worry and they are also likely to be more escape-proof than converted aquaria.

Start to handle your snake as soon as you get it. House Snakes are, in my opinion, well suited to handling as a pet, although they will bite if your fingers are wiggled about like a pinkie in front of their noses.

**FOODS AND FEEDING**

Feeding is rarely a problem as these are greedy snakes that always seem to be hungry or willing to feed. House Snakes are capable of taking quite large prey in relation to their head and neck size. They tend to be good feeders and rarely refuse food, or fast, which is not uncommon in other species. In general, hatchlings should be offered a couple of small pinkies every 3-4 days. Feed juveniles 2-3 pinkies or furries every 6-8 days and adults 1-3 adult mice every 10-14 days.

**BREEDING**

A female House Snake may lay between 8-15 eggs per clutch. Any discoloured or wrinkled eggs should be removed and the remaining ones incubated at 26.5°C (80°F). Eggs should hatch after approximately 60 days into babies measuring on average 20.5cm (8in) in length. Each hatchling can be fed a pinky after one week of birth, but make sure all the snakes obtain their fair share of food. Raise the young in small, escape-proof containers. Baby snakes can escape from the tiniest hole or curl up and hide in the smallest piece of cork and will be hard to spot by the untrained eye.

A juvenile House Snake should be purchased when it is at least 20-30cm (8-12in) long. It may initially be kept in a fairly small pet home, heated using a heater pad. Once it starts to grow, move it to a 60 x 38 x 38cm (24 x 15 x 15in) vivarium (this unit is suitable for up to 6-9 months) or straight into a 90 x 38 x 38cm (36 x 15 x 15in) vivarium that will still be large enough once your snake is fully grown.

# ROYAL PYTHON

*Python regius*

**In the wild these snakes may be spotted at dawn or dusk, hunting or waiting to ambush their favourite food – the 'jumping mouse', or gerbil. Once seized, their prey is suffocated and swallowed whole.**

Occurring in west and central African grasslands and forest clearings, Royal Pythons are small and thick-set, measuring on average 100cm (39in) in length, although a mature female may exceed 200cm (78in). This species is familiar with hobbyists because its size and docility make it a perfect vivarium pet. In captivity they prefer a dimly-lit setup and shy away from intense light.

▲ **This Royal is now the right size to start handling. They are smaller, more manageable and more suitable as pets than other pythons.**

## CREATING THE RIGHT ENVIRONMENT

For a single mature specimen a 120 x 45 x 45cm (48 x 18 x 18in) vivarium is a minimum requirement. It should be well-ventilated and have front-sliding doors for access. The temperature should be maintained at 24-29.5°C (75-85°F). Provide shelter in the form of logs, rocks or natural cork to create resting areas for the snake. Wood chips, leaf litter or plain paper can be used as substrate. Your snake will appreciate a good-sized water bowl.

Keep the vivarium relatively dry but mist spray periodically with warm water. This is especially helpful for the python when it is ready to slough its skin because the increased humidity helps to lubricate the old skin.

## HANDLING

Royal Pythons are usually very easy to handle. Their initial eagerness to curl up into a ball disappears once they accept that you pose no threat or danger. They are nervous and tend to back away from any initial contact, but will very rarely bite. In the USA they are known as Ball Pythons because of their hedgehog-like habit of curling into a ball shape when threatened. As a stout snake, their body weight should be supported at all times.

▲ **Passive and shy, Royal Pythons are a very inoffensive and popular species. They are solid looking, heavier and more rotund than many other species of snake.**

## AMAZING FACT

Pythons are regarded by some African tribes as the Spirit of Rivers, God of War and Goddess of Fertility, often being kept in shrines. At one time in the State of Dahomey sailors in the early days of the British Empire were put to death for harming them.

**FOODS AND FEEDING**

Royal Pythons have a well documented tendency to fast in captivity. There may be several reasons for this type of behaviour. It is more than likely that a recently moved or imported specimen will not eat until it has settled and is comfortable in its new home, so allow a few weeks for this acclimatization to take place.

A lack of appetite may also be due to the wrong foodstuffs being offered at the wrong time of the day. From experience, I have found that the easiest way to start these pythons feeding is to allow them approximately two weeks to settle in to their new environment and then early one evening (reduce lighting if necessary) offer a dead gerbil, or a brown mouse. If this is refused remove it after 30 minutes or so and try again a few days later. If your offering is not accepted, you may wish to offer another dead, but warm, specimen. In this way, an older snake can be weaned off live food. Of course, not all snakes will be so willing.

In captivity, the record held for a fasting Royal Python is probably 26 months, with only water being taken during this period. Should a well-fed Royal Python stop feeding for several weeks do not worry unduly – this seems to be perfectly normal behaviour. In due course, the snake will feed again.

A baby farmed python will usually accept warmed prey readily but if it needs encouragement, try wiggling the food about on long tweezers. I have known many Royal Pythons who feed with astonishing appetites on dead mice, rats, sparrows, pinkies and fuzzies, and on one occasion a piece of raw chicken started a stubborn individual feeding.

When ready to feed, your python may eat two or more mice per week. It is just as likely, however, to eat nothing for a couple of months. When cooling pythons for breeding, courtship and for gravid females, consumption of food can be virtually nil.

**◀ Squeezed within its coils a python's prey cannot breathe and suffocates quickly. Pythons are among the largest of all snakes. A Reticulated Python, *Python reticulatus*, is recorded to have consumed a bear weighing 91kg (200lb) and, measuring 9.9m (33ft), it is the world's longest snake.**

**▼ The two claws visible on either side of the vent are, in effect, the remains of rear legs from distant ancestors.**

## DEFENCE MECHANISM

When threatened, the majority of snakes will either flee or fight back. Most unusually for a snake, however, the Royal Python simply coils itself into a tight ball and hopes that its compact shape will deter any would-be predators.

### BREEDING

There are very few Royal Pythons that are bred in captivity. Most of the available specimens are either farmed or ranched in West African countries such as Nigeria or Ghana.

In the wild, pythons consume vast quantities of vermin and this helps to keep stored grain safe and dwellings rat-free. In return for this valuable assistance, the native peoples gather gravid (pregnant) females together and then harvest the young once they have hatched. Bringing in much needed money, to often very poor rural areas of Third World countries, this type of snake farming must be seen as a positive move and helps to reduce the need for taking snakes directly from the wild.

Males and females are distinguished by two claws visible on either side of the vent. These are in effect the remains of rear legs from distant ancestors – snakes are the last group of reptiles to have evolved. These claws are larger in males and are used to stimulate the female during courtship.

Mature, active Royal Pythons will mate and lay between 6-8 eggs above ground. These eggs are guarded and are partially incubated by one of the parent snakes. The parent remains coiled around the eggs until they hatch, usually about 100 days later if kept warm – about 30°C (86°F).

Newly-hatched Royals will rarely accept a feed until after their first slough (approximately 8 days after hatching). They are normally quick and eager feeders and possess a beautiful golden hue that is present only for the first few months of life.

**▼ Snakes have no eardrums. Instead, a vibration-transmitting bone is connected to the lower jaw. They are largely deaf to airborne sound but may 'feel' vibrations from the ground.**

# COMMON GARTER SNAKE

*Thamnophis sirtalis*

**This diurnal snake is native to North and Central America. Canadian Garters are often exported under licence as pets to the UK and many of them are now captive bred.**

In the wild, Garter Snakes are generally found near water and feed mainly on amphibians, including tadpoles, frogs, fish and earthworms. Many subspecies of this highly variable snake exist. Some of these inhabit drier regions and are more likely to take pinkies and mice than those from wetter habitats.

This is a slender and active species of snake that grows to 60-120cm (24-48in) in length. In captivity, they may live for 8-10 years – a natural life span in the wild would be considerably less.

## CREATING THE RIGHT ENVIRONMENT

A 90 x 38 x 38cm (36 x 15 x 15in) vivarium will suit an adult pair of Garter Snakes. They generally require a dry vivarium furnished with a humidity box and a large water bowl. Reptile grass or even paper can be used as a substrate, but regularly changed wood chips or larger stones are more attractive options. Logs, branches and other furnishings can provide an attractive home for a Garter Snake. The vivarium should offer plenty of hiding places and a good range of temperatures to allow your snake to thermoregulate. Keep temperatures in the range of 20-26°C (68-79°F) during the summer months and lower to 10°C (50°F) in winter. Due to the extremely long, cold winters of their natural Canadian environment, some Garter Snakes are only active for 3 months of the year.

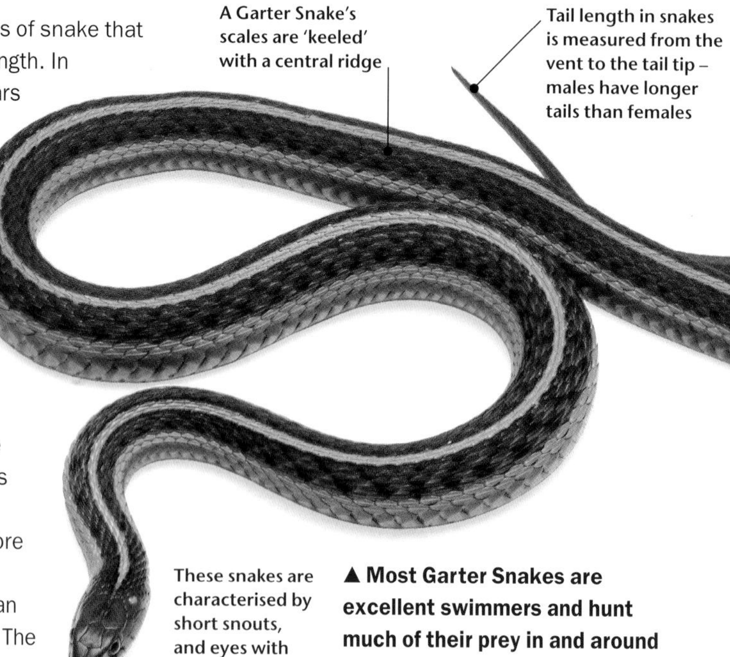


**▲ Most Garter Snakes are excellent swimmers and hunt much of their prey in and around ponds, streams and rivers.**

Follow a good hygiene regime. Keep substrates dry and clean to reduce the chance of harmful bacteria multiplying – heater pads are ideal for keeping substrates dry. Use a thermometer to check temperature and a thermostat is recommended to maintain an optimum temperature gradient.

### HANDLING

Garter Snakes handle well after initial training. Younger ones, especially juveniles raised in captivity, are most likely to remain calm when handled and can be fed carefully from the fingers or with long tweezers.

## Ribbon Snakes

*Thamnophis sp*

Ribbon and Garter Snakes look very similar and are members of the same genus. Ribbon Snakes have longer, more streamlined bodies and are able to cover open ground faster than Garters. Care conditions are virtually identical, but Ribbon Snakes will rarely eat earthworms, preferring treefrogs and other amphibians and fish for the bulk of its diet. Breeding is only likely if a group are kept together communally. Litters are small, with 4-5 young being born in early summer. The are rarely available from captive-bred stocks.

### FOODS AND FEEDING

Garters require more regular feeds than many other species of snake and should be fed every other day. Fish, fish pieces, earthworms, pinkies and small mice can be offered. A complete meal called 'garter grub' is also available.

It may be necessary to wriggle food around, especially to begin with, to stimulate a feeding response in your snake. It is well known that a diet of dead fish can cause a deficiency in thiamine (Vitamin B1), so fish pieces will need to be dusted with vitamin supplements. The processed 'garter' foods are supplemented to overcome this potential problem. Supplying a varied diet is the key to maintaining optimum health in this particular species.

### BREEDING

Garter snakes are live-bearers and the young grow rapidly if fed correctly, maturing at 8-10 months of age. The number of young can vary between subspecies and depends on the age, experience and health of the female. Between 10-80 young per litter is typical. Sperm is retained by females and they may produce young up to two years after one successful mating.

### AMAZING FACT

Garter Snakes rely greatly on their sight for hunting. As a diurnal predator they can track by sight and scent, seizing their prey firmly in the mouth. Prey is generally too small to need constriction and the majority of animals are swallowed head first.

# Invertebrates

We share our world with millions of invertebrates. They are the basis of many food chains and are the most abundant creatures on the planet. Predatory mantids and leaf-crunching stick insects are examples of these fascinating creatures.

Ninety seven percent of all living animals are invertebrates, that is, creatures that do not possess a backbone. These include lice, fleas, giant clams and squid. Some invertebrates have a special body structure called an exoskeleton, which is a hard external structure as seen on a scorpion or millipede. Other species, such as worms, are composed of soft tissues.

All invertebrates are ectotherms, which means that they are reliant on external sources of temperature to maintain their bodily functions. Interestingly, newly discovered species have been found to thrive under extreme environmental conditions, both above and below what scientists have conventionally termed fatal temperatures – for example, under ice caps and at thermal vents.

Many invertebrates, especially those that attack crops or invade homes, are in conflict with humans. For example, millions of pounds are spent on researching new methods of control for malaria-carrying mosquitoes, *Anopheles sp* and the potato-attacking Colorado beetle, *Leptinotarsa decemlineata*. Other species, such as the honey bee, *Apis sp*, are beneficial since they produce tons of honey and wax, and pollinate fruit trees.

Invertebrates consume a variety of foodstuffs. Some species, such as spiders and scorpions, are carnivores, while others, such as stick and leaf insects, are herbivores. Cockroaches are scavengers, willing to consume anything on offer. Foodstuffs are obtained in a variety of ways and may be grazed, hunted, ambushed, trapped, snared or lured, depending on the individual species of invertebrate.

Amongst the thousands of known species there is an enormous range of reproductive strategies to ensure the continued survival of this diverse group of animals. Many species produce eggs after completing a mating embrace, some are parthenogenic (unmated females producing fertile eggs) and most molluscs, including snails, are hermaphrodites.

In general, invertebrates are the least handleable of all the exotic creatures in this book. If you are in any doubt, do not handle them. They should not be moved or handled during a moult, since they are extremely vulnerable to damage at this particular stage. Soon after a moult an invertebrate's body hardens and they can then be handled if necessary. Adult supervision is essential when young pet keepers wish to move or handle these creatures.

▶ **Praying mantids are short-lived predatory invertebrates found in Europe, the Americas and most tropical regions of the world.**

Only a few invertebrate species are suitable to keep as household pets. Some are familiar to us all – insects, for example, which have six legs, a body split into three sections and an exoskeleton. These are the most numerous of all creatures on the planet – a large swarm of a single insect species, such as locusts, may contain more individual insects than the entire human population.

Some invertebrates are very attractive and, although somewhat unusual, are fast becoming desirable pets. Insects generally require little space (stick and leaf insects, for example) and are easy to maintain. They have fascinating life cycles and behaviour patterns that make them absorbing and educational pets.

**▲ The spider's internal structure is visible in this recently shed tarantula skin. Adults moult once or twice a year to enable them to grow and renew lost or damaged limbs. Spiders are frequently very hungry and thirsty after a moult.**

## INVERTEBRATE CLASSIFICATION (EXTRACT)

| Class | Description | No. of Species |
|---|---|---|
| INSECTA | Invertebrates in this class typically have three pairs of legs and are the only invertebrates to have wings as adults. Praying mantids belong to the order DICTYOPTERA and over 6000 species are known to man. Stick and leaf insects belong to the order PHASMIDA and there are approximately 2000 known species. | approximately 800,000 |
| ARACHNID | These creatures have four pairs of legs and the body is normally made up of two parts, with a thin 'waist' separating the cephalothorax (head) and the abdomen (body). Tarantulas belong to the sub-order ORTHOGNATHA and scorpions belong to the order SCORPIONIDAE. | approximately 70,000 |
| MYRIAPODA | Millipedes and centipedes belong to this group of animals. Millipedes are inoffensive scavengers. Their legs are contained under the body, which is often black, brown or delicately striped. | approximately 11,000 |

# INDIAN STICK INSECT

*Carausius morosus*

**Although still considered exotic, this species has been kept in the UK for over 100 years, and has been popular in schools and with pet keepers for almost as long.**

Indian Stick Insects are one of the easiest species of invertebrate to manage. Their bodies range from green to brown in colour and they possess bright red spots on the inside of their forelegs that are flashed as a warning when they are disturbed. If threatened, they will often drop to the ground and remain motionless like a twig. When the danger has passed they will open out their legs and start to move again. Once adult, female stick insects measure up to 11cm (4.5in). Males are rarely encountered. Life expectancy is approximately one year.

**CREATING THE RIGHT ENVIRONMENT**

Requiring a tall vivarium, measuring at least 30 x 45 x 30cm (12 x 18 x 12in), these insects will live mostly on their food plant. These should be potted up and changed regularly to ensure an abundant food supply. Fresh cut sprigs of privet or leafy lengths of bramble can be used, but unpotted plants will dry quickly and be of little benefit to your pets unless replaced often.

Keeping your insects at a temperature of 24°C (75°F) is ideal, but they are very tolerant and will be happy in a warm room. A reasonable level of humidity is required – regularly mist spray but ensure good ventilation to prevent moulds from growing. Thirsty stick insects exercise their mouth parts continuously and a quick spray of droplets onto a leaf soon quenches their thirst. Cover the floor of the unit in paper so that any eggs that have been laid are readily visible. Cleaning is minimal – just change the paper weekly.

### Giant Spiny Stick Insect

*Eurycantha calcarata*

A large chocolate brown species, with females growing to 15cm (6in) and males to 10cm (4in), they are well armoured, especially the males who are equipped with two large, sharp spurs on the legs to fight off rivals. Only females have an ovipositor. The nymphs vary from orange to green in colour. This species spends most of its time resting, often in large numbers, in logs or behind tree bark. More ground space is therefore required and a well-ventilated 60cm (24in) cubed vivarium is ideal. This species loves water and will drink quite often. A shallow saucerful should be offered on a regular basis. Handle your pet with great care – lift from underneath using the palm of your hand.

## HANDLING

These little creatures tame easily but need to be handled with great care. Gently pick up your insect between finger and thumb, or offer a flat palm, and tap it so that it walks onto your hand. Legs that are stuck or hooked onto an object or piece of clothing need a gentle nudge to loosen their grip.

A classic insect, with six legs and a body divided into three parts: the head, thorax and abdomen

Antennae are covered in tiny sensory hairs to help the insect understand its environment

Privet is an easy plant to grow and can be propagated into small pots suitable for the vivarium

## FOODS AND FEEDING

They are herbivorous by nature, eating only plant material throughout their lives. Privet and bramble are their favourite foods. You can plant these in your garden at little or no cost and provide your pet with a lifetime's supply of food. However, make sure you do not release the insects into your garden.

## BREEDING

Stick insects are parthenogenetic, capable of reproducing without mating. Adult female Indian Stick Insects lay fertilized eggs measuring about 2mm (0.08in). Hundreds of these tiny eggs will be laid during her lifetime, several a day during peak output. These smooth round 'pods' are easily separated from droppings if further young are required. Place them on sand or vermiculite in a clear tub and leave in a warm place. Remember to check them periodically.

The young can take up to a year to hatch and require identical vivarium conditions to adults, but they must not be overcrowded because they will eat each others' legs. As babies appear, place them in a rearing container supplied with plenty of new privet for them to feast on.

A baby stick insect, or nymph, will grow to adulthood within a few months by moulting six times. Whilst moulting they will suspend themselves from vegetation or the container lid. During this stage they are extremely vulnerable and should be left alone to drop out of their old skin and harden their new one before handling. The old skin will usually be eaten.

**◀ Many species of stick insect can be kept under similar conditions. Simple to keep, they are a good introduction to keeping invertebrates.**

# LEAF INSECTS

*Phyllium sp*

**Their bodies are generally flat and green. Brown 'tatty' edges and the odd hole complete their camouflage and ensure they look exactly like the leaves upon which they live and feed.**

The Leaf Insect is native to the warm and humid Malaysian jungles and, depending on the species, will measure 8-15cm (3-6in) once adult. It is shorter lived (6-8 months) and more difficult to maintain and breed than stick insects, but it is undoubtedly worth the extra effort.

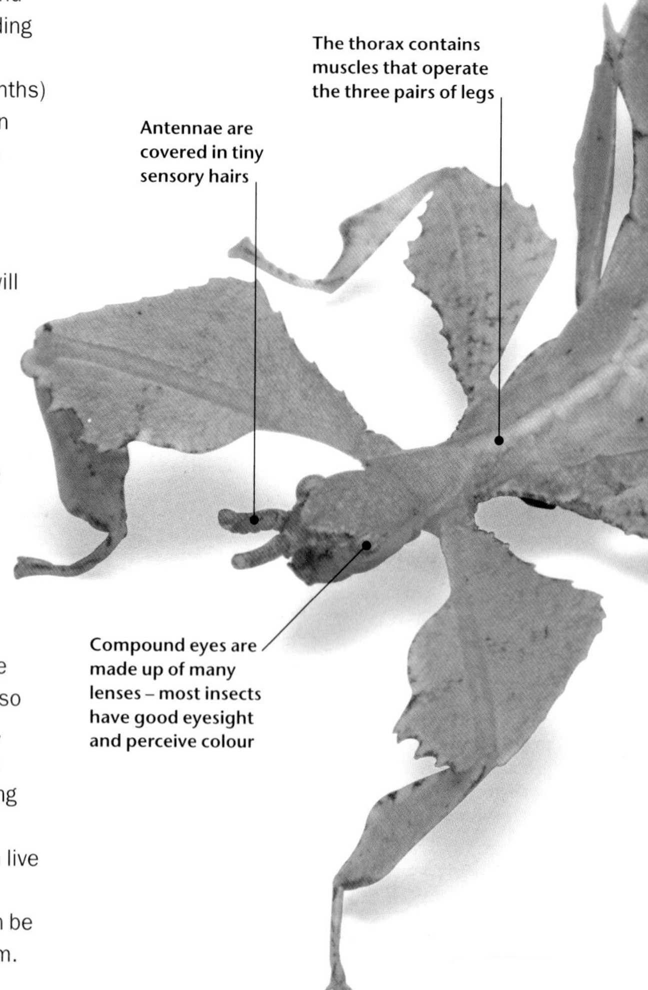


**CREATING THE RIGHT ENVIRONMENT**

A 60 x 45 x 30cm (24 x 18 x 12in) vivarium will house up to 6 individuals. Whilst preferring temperatures in the region of 21-27°C (70-80°F), they do not like stagnant air, so good ventilation is essential for optimum growth and development, especially during a moult. Maintain the temperature within the vivarium using heater pads or keep it in the warmest part of the home.

Since leaf insects are arboreal creatures the choice of substrate is largely decorative. However, plain paper should be placed on the base of the vivarium when they are breeding so that eggs may easily be located. Wood chips, or even moss, kept moist would help to keep humidity at around 70%. Regular mist spraying with tepid water and use of a hygrometer is recommended. In captivity, Leaf Insects both live and feed on bramble leaves. Old sprigs need replacing every other day, or small plants can be potted up and placed directly into the vivarium.

▲ **Leaf Insects are worth the effort that is required to keep and breed a colony in captivity. This sub-adult is almost ready for breeding.**

▲ **Insects excel at camouflage and the detailed design and colour of the Leaf Insect would make spotting one in the wild a remarkable achievement.**

## HANDLING

Adults may be encouraged to walk onto your hand. They wobble about in a highly comical fashion trying to convince you that they are a leaf blowing in the breeze. Handle them with care because it is so easy to damage these very delicate creatures.

## FEEDING

Leaf Insects are herbivorous. They eat Guava and other tropical leaves in the jungle but, in captivity, they will accept bramble leaves as a substitute. Fresh sprigs should be inserted into the vivarium every other day or so; alternatively, pot up a few small plants.

## BREEDING

These small invertebrates have a life expectancy measured in months, not years, and at about 4-5 months of age they are ready to mate. Males are easily identified by their slimmer bodies and longer wings.

About 100 eggs are produced singly. These drop to the ground and are camouflaged to look like the droppings of a silk moth caterpillar living in the same region in the wild. A few weeks after the eggs are laid the parents die. Incubate the eggs in the same way as Stick Insect eggs.The babies are bright red and will immediately climb up into bushes and start feeding on leaves.

# GIANT MILLIPEDES

*Epibolus sp*

**There are about 8,000 recorded species of millipede. Commonly called 'jungle trains' or 'chongaloloo' in Africa, several different varieties are now available to the pet keeper.**

These are essentially animals of soil and leaf litter. They are easy to establish and it is quite possible to breed them providing several are kept together as a colony. They may be reddish-brown or black, and some species are even striped. Growing to 25cm (10in) in length, they can live for several years if they are kept in the right conditions. Millipedes are slow, harmless scavengers and should not be confused with centipedes.

## CREATING THE RIGHT ENVIRONMENT

A large container is necessary. One measuring either 90 x 45 x 30cm (36 x 18 x 12in) or 60 x 30 x 30cm (24 x 12 x 12in) will make a suitable home. Add chemical-free compost, or peat, to a depth of 6-8cm (2.5-3in) and cover with a quantity of leaf litter – oak, sycamore and moss are ideal. It is advisable to pre-freeze leaf litter to kill unwanted mini-beasts that may colonize your millipede's home. Logs and small pieces of cork chip are used to cover the soil and provide grazing. Maintain a temperature of 24-27°C (75-80°F) using a heater pad. Mist-spray occasionally to keep humid, but make sure it is not too damp or wet.

You only need to clean out your vivarium 1-2 times a year. If you keep a large number of millipedes together, you will need to do this more often. All compost should be disposed of and replaced.

## HANDLING

Millipedes are active at night or under crepuscular (low light) conditions. Many species can secrete unpleasant fluids from glands along the sides of the body when disturbed or offended. These may smell, sting or stain clothing. When disturbed they will coil up tightly for protection. They will also curl up when at rest or sleeping.

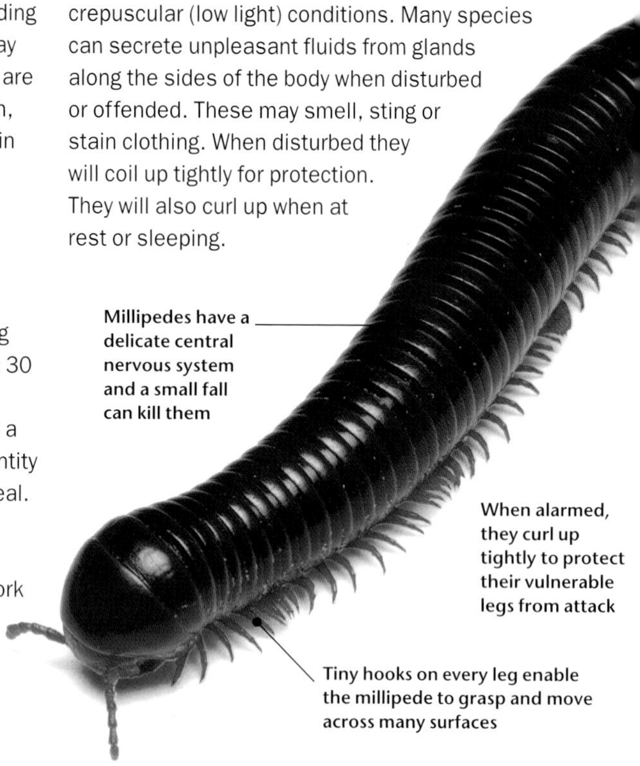

Millipedes have a delicate central nervous system and a small fall can kill them

When alarmed, they curl up tightly to protect their vulnerable legs from attack

Tiny hooks on every leg enable the millipede to grasp and move across many surfaces

Frequently handled specimens stop secreting these fluids because they know you will not harm them. Some people do wear gloves, but the main thing is to treat these invertebrates with respect and learn to enjoy their company.

▼ **The name millipede, meaning 'thousand feet', is an exaggeration. No species has many more than 300 legs.**

## Florida Millipede

*Chicobolus spinigerus*

This 8-13cm (3-5in) species is attractively marked and found in south-eastern USA, namely from Florida and northwards into South Carolina. They should be kept at temperatures of 21-27°C (70-80°F) in the same conditions as Giant Millipedes. Numerous native American species of millipede exist, but few are as large as those imported from the tropics. Temperate countries like Great Britain have tiny millipedes rarely exceeding a few centimetres in length.

### FOODS AND FEEDING

Besides eating leaf litter, small cubed pieces of vegetable and fruit may be offered to your pet on upturned jam-jar lids or small saucers. These should be removed every few days, and you will soon get an idea of what your particular millipede likes. Millipede waste appears in the form of small faecal pellets and it can be removed occasionally from the vivarium; it makes excellent compost for plants. As the leaf litter in their home disappears it will need to be replenished.

### BREEDING

Males are easily sexed when viewed from underneath – gonopods are visible on the seventh or eighth segment. These are modified legs used to transfer sperm to the females. In a male millipede these modified legs are frequently hidden from view.

Millipedes wrap around each other to mate. Some then build a nest in which to lay eggs. Should yours breed, you will soon notice numerous babies crawling about the leaf litter. Under a magnifying glass these can be seen to have much fewer segments and legs. There is no need to move them to a separate rearing container since they will be quite safe with Mum and Dad, soon developing more legs.

▲ **Each measuring only 6cm (2.5in) in length, this trio of young captive-bred millipedes are ready to start a new colony.**

# PRAYING MANTIS

*Sphodromantis sp*

**The first part of its name describes well the mantis' famous pose, with forelimbs raised up to its head as if in prayer. It is also one of the few insects that can turn its head and 'look over its shoulder'.**

There are about 2,000 species of Praying Mantis worldwide; none are native to the UK. Most of them are found in the tropics but some are native to Europe. At least one European species, *Mantis religiosa*, was introduced into the USA early in the twentieth century by accident, in plant crates, and is now widespread and considered a naturalized species. Several are available from pet stores but many more through membership of relevant specialist societies.

Generally green or brown in colour, one particularly attractive species, the Orchid Praying Mantis, *Hymenopus coronatus*, is a shade of pink closely resembling the flower blooms it rests upon whilst awaiting insect visitors.

The Praying Mantis is a voracious predator. It will actively stalk its prey or else wait patiently, camouflaged against foliage, for an insect to venture within striking range. Upon close examination we can see that mantids are well designed for killing. The long front legs are lined with numerous spines and a curved hook which, in a fraction of a second, snatches and impales the unsuspecting insect. Once held, the prey's fate is sealed and the mantis starts to consume them alive.

Your pet mantis will usually grow to 7.5cm (3in) in length and will live for approximately one year.

## CREATING THE RIGHT ENVIRONMENT

Praying Mantids are generally purchased as nymphs, individually packed in small plastic containers covered with netting or punctured with tiny breathing holes. Measuring only about 2cm (0.75in) in length, your mantis will reside on a twig or other object upon which it can climb. As long as it is fed on fruit flies or other tiny prey and is kept warm at 24°C (75°F), this small space will be sufficient for a while. As the mantis feeds and

▲ **Young mantids are reared separately in small containers. Simple to keep and with a life span of one year, they are a short-term commitment.**

needs to grow it will moult, reaching its next development stage (known as an instar). Having moulted several times, larger accommodation should be offered.

After 5-6 moults, your Praying Mantis will reach adulthood. Probably measuring no more than 7.5cm (3in), a 60 x 45 x 30cm (24 x 18 x 12in) vivarium or a tall unit measuring 30 x 45 x 30cm (12 x 18 x 12in) will be ideal. Anything larger than this and you might not be able to see and enjoy your pet, so effective is their camouflage. A daytime temperature of 24°C (75°F) should be maintained – a heater pad will do this for you. This can drop by a few degrees at night with no ill effects. Housing should be well-ventilated. Mantids should be kept individually because they are just as happy to eat one another as to eat their normal prey.

As arboreal invertebrates, mantids need to be provided with something to climb upon.Branches, twigs, foliage and other decoration will provide this and make an attractive home for you and your mantis to enjoy. A living plant really helps to enhance a display, but do ensure that the light bulb is not positioned too close to it. Perhaps consider using a full spectrum fluorescent light instead to maintain plant health.

**▼ A voracious predator, the Praying Mantis is a glutton, forever on the look out for unwary prey.**

▲ **Praying Mantids are usually green like this attractive specimen, but some varieties mimic colourful flowers or even dead leaves found on the forest floor.**

## HANDLING

For close examination you can encourage a mantis onto the palm of your hand, where it will move around from hand to hand quite happily. Don't attempt to pick one up using your finger and thumb because the mantis will jab you with its forearms. Mantids can fly, so ensure windows and doors are closed before allowing a mantis to climb on your hand.

## AMAZING FACT

Many Kung Fu moves are derived from the movements of the Praying Mantis. 400 years ago Wang Lang, already a martial arts expert, was inspired whilst watching a mantis kill a cicada. He developed his style using a refined series of moves mimicking the mantis' ability to strike and grab.

## INSTARS & MOULTING

An instar is a development stage, and the number it takes to reach adulthood depends on the species. Moulting is the process between instars enabling growth and development to take place.

## FOODS AND FEEDING

Mantids will eat almost any live foods that it can catch and hold in its powerful forelegs. The youngest will eat fruit flies, but once they begin to grow they will quickly graduate to locust hoppers, crickets, butterflies, grasshoppers and flies. In the wild, even bees and wasps will be eaten. Feed your pet as much as it will eat. Leave a few suitably-sized prey insects in the vivarium so that your mantis can feed whenever it is hungry.

They are messy feeders, leaving discarded legs, wings and other body parts lying around the vivarium floor. These should be removed on a regular basis and, apart from removing faecal pellets, this is the only cleaning required.

Your mantis may drink the odd drop of water direct from leaves and should be sprayed with water every day or so. Make sure that your vivarium is airy and well ventilated.

## BREEDING

Mantids are essentially an annual species, with offspring emerging each spring. If lucky, they survive to mature, mate and, if female, lay eggs before dying within the same year. Your pet mantis is therefore relatively short lived, taking six months to mature and then living for approximately six months as an adult. By growing on several nymphs to maturity, you have a good chance of breeding them.

▲ **Baby mantids are protected before hatching by the hardened foam exterior of their egg case.**

Females are easily distinguished from males. They have six visible segments on the underside of their abdomen, whilst males have eight. Mating is a dangerous time for males – they may be eaten attempting to mate, during mating or immediately afterwards. One female may produce several egg cases, or oothecae, from a successful mating. Even unmated females will lay eggs, but these will not hatch.

The new generation emerge in great numbers from each egg case, which may contain 20-200 eggs. Newly-emerged mantids are the size of a mosquito and are miniature versions of the adults. Like all insects, by the time they have reached adulthood they have wings and can fly. The appearance of wings establishes that maturity has been reached and that the mantids are ready to breed.

# TARANTULAS

*Family: Theraphosidae*

**There are approximately 800 species of Tarantula. They are large and hairy, often possessing interesting colours and markings. Tarantulas are predators and have large fangs.**

No record can be found of human deaths caused by a tarantula bite – they are not considered a species dangerous to man. Tarantulas are popular and can be very low cost pets, requiring only the minimum of time and space to maintain in excellent condition. Bred in the UK in large quantities, it is recommended that you purchase one of the test-tube reared spiderlings measuring between 1-2 cm (0.5-0.75in). The most popular species bred and kept in captivity come from North, Central and South America. One of these is the Curly-haired Tarantula, *Brachypelma albopilosa*, my personal favourite ('Erica' is shown below).

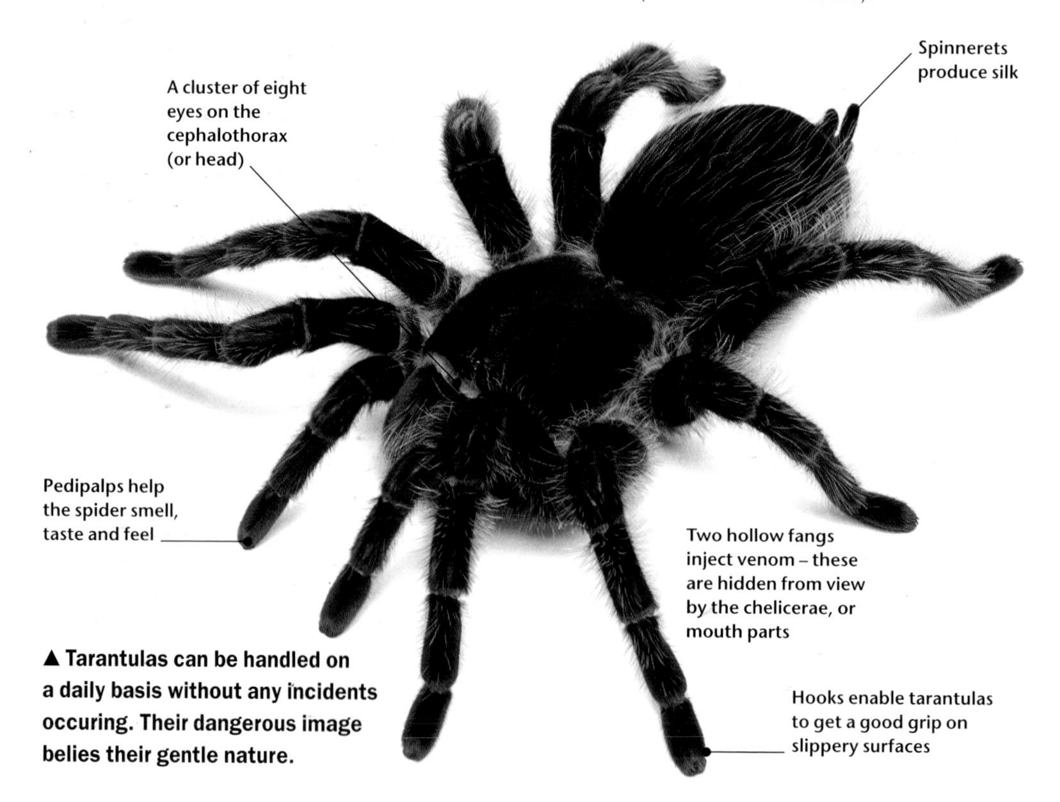


**▲ Tarantulas can be handled on a daily basis without any incidents occuring. Their dangerous image belies their gentle nature.**

▲ **This double spider unit safely separates two adult tarantulas.**

**CREATING THE RIGHT ENVIRONMENT**

Most species of tarantula should be kept singly in glass containers. House spiderlings in small jars or plastic containers and increase the container size as the spider grows. A 30 x 30 x 20cm (12 x 12 x 8in) unit is sufficient to house a single adult for its entire life. Although a large vivarium measuring 60 x 30 x 30cm (24 x 12 x 12in) or 90 x 30 x 30cm (36 x 12 x 12in) may be used to create a more natural-looking landscape.

Care must be taken with larger units, however, to ensure that a high enough humidity level is maintained. The experience of numerous tarantula breeders and keepers shows us that smaller units appear to be more beneficial to most species – they often spend days, or even weeks, squeezed into 'burrows' the size of empty toilet rolls within the vivarium. A hygrometer will help you to measure humidity levels accurately. Desert species of tarantula require 50%-70% relative humidity and tropical species need 70% - 90% relative humidity.

## Chilean Rose Tarantula

*Grammostola cala*

This popular species has a salmon-pink hue and is one of the best spiders to keep with a view to occasional handling – experienced adult supervision is required. Like most species of tarantula it should be kept singly, except for breeding purposes.

Glass spider tanks are available and these help to keep the humidity levels high. Fine ventilation mesh ensures that the air does not stagnate. These glass units are preferable to over-ventilated plastic pet homes. The majority of species are ground dwelling and they require a good quality substrate and a cork 'cave', burrow or similar shelter. The base should be covered with substrate to a depth of 8-10cm (3-4in); peat, vermiculite or a mixture of both should be thoroughly dampened to provide ideal conditions. A cork log or similar shelter should be pushed into the medium or rested on top of the substrate. Maintain the temperature at 24-29°C (75°F-84°F).

A unit containing a solitary tarantula requires virtually no cleaning – maybe only once a year. When cleaning out the vivarium, it is best to 'usher' your spider into an escape-proof pot or tub for easy and safe removal. The vermiculite or other substrate should be disposed of and cork or other loose objects cleaned thoroughly. Never use disinfectants, insecticides or bleach; boiling water will be sufficient. Freezing cork or logs is another way to minimize the risk of introducing unwanted pests into your vivarium.

**HANDLING**

The best advice I can give is never to handle your pet spider. Many people, however, wish to experience the thrill of such intimacy. If you are intent on trying to hold one, the following information should be taken into account. Handling a tarantula should only be attempted when an experienced adult is on hand to assist. These spiders have large fangs which will definitely hurt if you are bitten and their venom is similar in strength to that of a bee sting. Many people are allergic to venoms, and bee venom kills more people annually than any other venomous creature. Some species of tarantula, such as the Curly-haired or Chilean Rose, seem to be the most passive and friendly in my experience. Handling any spider is not recommended for people of a nervous disposition.

Never handle your pet during, just before or after a moult, or just after feeding. Start handling your spider when it is still young by gently encouraging it to walk onto your hand from its vivarium, or from a flat surface such as a low table. You should be sitting down on the floor, so that if you get scared and drop the spider, or the spider gets scared and jumps off you, serious injury from a fall onto a hard surface can be avoided. Do not put your tarantula close to your face or eyes just in case it rubs off some hairs, which can cause irritation.

As soon as a tarantula appears agitated or 'rubs hairs', stop handling it. A tarantula may defecate on you; it does not hurt, but can give some people a shock. Don't let your tarantula crawl onto clothing because the tiny claws on each leg may get caught and stuck in the fibres. Never handle Asian Bird-eating Tarantulas or specimens of the *Poecilotheria sp.* A person was once hospitalized after a bite from an unidentified *Poecilotheria* species in Sri Lanka.

**▲ People get a thrill from handling tarantulas. However, such close contact is best left for experienced individuals to teach the uninitiated.**

## FOODS AND FEEDING

All species of tarantula, ranging from juveniles to adults, can be reared on insects but the size of prey should reflect the capacity of the spider. Make sure you offer a variety of food types.

Young tarantulas (spiderlings) will feed on hatchling crickets or fruit flies (especially flightless ones), and as the spiderling grows and moults larger prey should be offered. On average, foods should be approximately a quarter to one-third the length of the spider. You will soon see, if observant, what size or type of prey is attractive to your pet.

### Pink-toed Tarantula

*Avicularia avicularia*

This non-agressive arboreal species is a fast moving and very attractive fuzzy black spider with pink 'toes'. A 60cm (24in) cubed vivarium would make a lovely display unit for 4-5 of these 'social' tarantulas. Spiders of similar size may be kept together and cork logs, branches or other climbable structures should be included in the setup. These spiders require 70%-80% humidity and a temperature range of 24-28°C (75-82°F) They build hammock-like webs, creating an interesting visual display. When several inhabitants are living together, lots of 'action' is likely to be observed. This is less likely with a solitary specimen, especially if it is a terrestrial species that hides in a burrow for most of the day.

Crickets, mealworms, maggots, moths and house flies are all potential food sources. Arboreal species of spider will prefer flying or climbing foods, whereas ground dwellers will prefer crickets, terrestrial bugs and caterpillars. Offer your pet foods on a regular basis, at least twice a week to juveniles and every 9-11 days to more mature animals.

### MOULTING

When your tarantula is found upside down or on its side in the vivarium, it is probably moulting. Do not disturb it and carefully remove any live food still present. It can take between 1-2 days for an adult to complete the process. Spiderlings moult very regularly because they grow quickly. Adults do so less frequently and moult on average once a year.

Care must be taken to prevent the spider's unit from being overrun with uneaten live foods. These may annoy your tarantula or cause it terrible damage and stress if it begins to moult. Tarantulas refuse food prior to a moult, but are generally ravenous soon after. Long tweezers will help to remove any carcasses or uneaten prey without risking damage to your fingers. A well-fed tarantula, with a rounded abdomen, may only require infrequent feeds and will often eat nothing for weeks, or even months, at a time.

Spider venom is a specially modified digestive enzyme similar to, but more potent than, our own saliva. Its main use is to capture, kill and pre-digest food; its defensive role is secondary.

In the wild most spiders obtain all the water they need from their humid environment and from food. In captivity it is a good idea, however, to have water on offer at all times.

**BREEDING**

Breeding tarantulas provides a fascinating insight into the precarious mating habits of spiders. You are, if successful, likely to feel euphoric at your achievement in helping to bring new lives into the world. However, before embarking upon this large venture, you might wish to consider whether you have a ready outlet for approximately 400-600 spiderlings and adequate time and energy to ensure their survival and development.

Initially, you will need to sex your spiders and to ensure that they are well fed, in particular the females. Male tarantulas mature after a particular moult; mating hooks will be visible near the joints on the underside of his front pair of legs. Once a male spider has matured, its life span can be measured in months. The odd individual may survive for a year. Female tarantulas are less easy to identify, but they tend to appear to be heavier and more rounded and have a larger abdomen.

**CAUTION**

**An unhappy or threatened tarantula will release irritating hairs by rubbing its back legs against its abdomen (only New World species). This cloud of hairs floats into the air and onto the skin or face of their aggressor, causing a painful rash or temporary blindness.**

**▲ This Asian Bird-eating Tarantula is very aggressive and has reared up into a threat posture. Any attempt to touch it would result in a bite. Spiders of the family Theraphosidae are usually more friendly.**

Mating carries considerable risks to male tarantulas – many are killed or badly injured in their attempt to mate with the longer-lived females. A male that is ready to mate will have deposited some sperm onto a small sheet of web. He will then dip his pedipalps into the sperm and proceed to visit a potential mate; a female releases an attractant scent that causes a male to hunt her out.

If the male finds a willing partner, he will hold her fangs with his mating hooks and insert, and empty, his palps into her genital opening. You may, if things don't go to plan, need to leap in and separate them if fighting occurs. A plastic pet home is ideal to drop over each specimen without putting your fingers at risk. Once mating has taken place the male will retreat hastily to avoid becoming a post-nuptial meal.

Several days later the female will lay her eggs. Numbering into the hundreds, these are contained within a silk bag and protected by the mother until they hatch 8-10 weeks later into tiny colourless replicas of their parents. At this time, seperate each spiderling into tiny rearing containers and feed on small live prey such as fruit flies. As they grow, they gradually aquire the markings of their parents. The parents, if both have survived, ought to be fed well to rebuild their strength if another mating attempt is to be made.

## ASIAN BIRD EATERS

Native to Thailand and south-east Asia, Asian Bird Eaters (a specimen is shown opposite) have attractive velvet markings on their bodies and lack the irritating hairs possessed by New World species of tarantula. Despite their name, these spiders mainly eat insects, such as crickets. They are rarely captive bred and are fast moving, very aggressive and eager to bite – do not attempt to handle. This species is only suitable for the experienced tarantula keeper.

## Mexican Red-kneed Tarantula

*Brachypelma smithi*

These are one of the most striking of all tarantulas – large in size and, as their name suggests, with bright red knees. This spider is a protected and threatened species. It is more expensive than other spiders and is available only from captive-bred stocks. The females can be long lived – one has been recorded alive at 30 years of age.

I have in my collection four adult specimens on loan from HM Customs and Excise. They were part of a seized shipment from Mexico, imported without the correct paperwork. Each spider lives in a unit measuring 20 x 20 x 20cm (8 x 8 x 8in) and I have had them for seven years. Ideal living conditions would consist of a 30 x 30 x 30cm (12 x 12 x 12in) vivarium, maintained at temperatures of 24-28°C (75-82°F) and 70% relative humidity.

This species is traditionally used for human contact at many zoos. However, I have found some individuals are easily annoyed and they will rub off their abdominal hairs at the slightest provocation.

# IMPERIAL SCORPION

*Pandinus imperator*

**These creatures are native to the humid, tropical forests of central Africa and spend much of their time under leaf litter and in the cracks found in tree bark and fallen logs.**

The largest scorpion in the world, measuring 15cm (6in), may seem an unsuitable pet. However, it is quite easy to care for and it is not considered dangerous to humans. Living up to 8 years, the Imperial Scorpion is occasionally bred in captivity but is mainly imported from Africa. This species is covered under CITES.

**CREATING THE RIGHT ENVIRONMENT**

This forest scorpion, lives in burrows dug out under rocks, logs or other sturdy objects. Much of its time will be spent beneath ground, only venturing out after dark to hunt for food. Peat or soil, 13-15cm (5-6in) deep, are ideal substrates to place in the vivarium. Some parts should be

▲ **Beauty or beast – scorpions are unusual pets that are easy to maintain in captivity.**

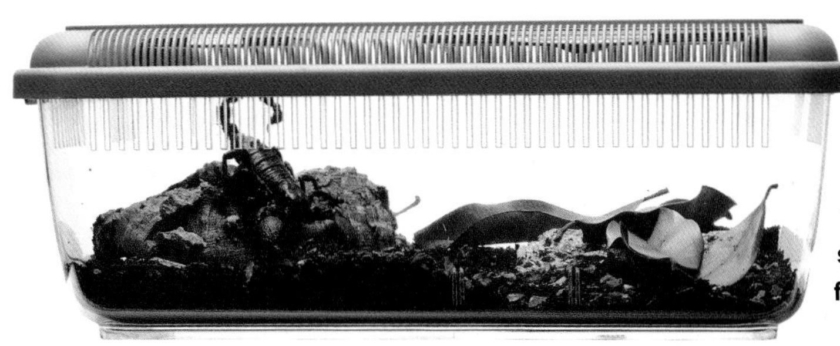

◀ **Compost, leaf litter and cork bark, added to a depth of 13-15cm (5-6in), provide the correct environment for this scorpion. The acrylic vivarium is ideal – the lid fits securely and prevents your pet from escaping.**

covered with wood, logs and rocks but keep at least part of the ground clear so that the scorpion can move about freely.

Maintain the temperature at 25-30°C (77-86°F). Spray regularly to maintain humidity and to provide an additional water source for these creatures, since they will also drink from drops on plastic plants or wood. Ground warmth is important for scorpions and a heater pad should be placed under one third of the vivarium. A 60 x 30 x 30cm (24 x 12 x 12in) vivarium is fine for a single specimen.

A small colony of 4-10 individuals may be housed together for breeding purposes, so long as ample retreats and food are available. A vivarium measuring 120 x 30 x 30cm (48 x 12 x 12in) is adequate. Some fighting is inevitable within a small colony, and occasionally a specimen may be devoured by its 'friends'. For this reason many pet keepers prefer to keep specimens separately.

## HANDLING

I do not recommend that novices or young people handle Scorpions. If you want to try, hold your pet firmly by its sting and lift quickly onto your palm. This should only be done under experienced adult supervision. I hold my black scorpion regularly and have had no stings. However, the occasional pinch is to be expected. Beware though, some people are highly sensitive to animal venoms.

## FOODS AND FEEDING

As predatory arachnids, these creatures prefer slow-moving prey such as large grubs and caterpillars. Dead mice, pinkies and locusts will also be accepted. Crickets, however, are generally too fast for the Imperial Scorpion to catch and may eventually establish themselves in the vivarium. Offer food twice weekly by releasing insects into the vivarium or by offering food, using long tweezers, directly to the scorpion.

## BREEDING

Equipped with a pair of savage pincers and a poison gland in her tail, a potential female mate is approached by an amorous male with great caution. He swiftly grabs her pincers with his and, with her weapons safely checked, they start to dance, shuffling forwards and backwards. The male deposits a pack of sperm on the ground and as he pulls the female over it she takes up the pack into her body. To avoid becoming her supper, the male quickly relaxes his hold and they separate.

Eventually, some 20-50 live young appear as small white spots crawling on the mother's back. Within two weeks, once they have had their first moult and their exoskeletons have hardened, the babies leave the mother's back to fend for themselves. Maturing Imperial Scorpions shed their exoskeleton about once a year.

# Preventing and Recognizing Exotic Pet Ailments

Prevention is the key. Most health problems can be prevented, since the major cause of herptile ill health is poor husbandry. A lack of proper hygiene and/or a bad feeding regime are largely responsible. By providing the correct environment with the right quantity and quality of foods most problems can be easily prevented.

I recommend that the novice purchases captive-bred specimens wherever possible. Imported, wild-caught species are more likely to suffer from internal and external parasites and bacterial infections. Many of these problems can be overcome with correct medication and laboratory style/quarantine housing. If you think you have problems of this nature, contact your specialist society or dealer for advice. Also, if you think your pet is sick or suffering, contact your veterinarian or herptile veterinary specialist immediately. Most ailments that occur can be attributed to the environment, diet or injury caused by you or another animal.

## OBESITY

One of the major general health problems of all pets is obesity. Too much food or a diet too high in preferred or fatty foods can cause serious problems. It is impossible to state exactly what quantity of each foodstuff should be given to any species, but try to imagine life in the wild where most species would not eat every day or all year round. Encourage activity in predatory herptiles by scattering food – this makes your animal 'work' for its meal. This type of environmental enrichment is popular now in zoos. Dragging a defrosted mouse around the vivarium using long tweezers for a snake or releasing some bugs into the unit for a Leopard Gecko to physically stalk will aid this type of positive activity. Such 'exercise' will help prevent your pet from becoming a 'couch potato'.

▲ **Obesity is one of the largest causes of ill health in pet animals – always feed in moderation.**

## VITAMIN AND MINERAL SUPPLEMENTS

These are essential for many species but correct dosage is necessary since a lack, or excess, of vitamins and minerals can lead to problems. They are of particular importance for juveniles which are developing quickly. Follow manufacturers' guidelines for the duration of your pet's life.

## METABOLIC BONE DISEASE

This disease is very common in some species of captive-bred reptiles. Without the exposure to UVB rays they cannot synthesize Vitamin D3 in their bodies. Reptiles need Vitamin D3 to absorb calcium to form healthy bones. There are a variety of products available and I would advise a fluorescent full-spectrum tube, with both UVA and UVB, or giving a powdered Vitamin D3 supplement.

## ABSCESSES

Common in both snakes and lizards, an abscess may appear as a raised bump underneath the scales. Veterinarian treatment is required to surgically remove the infected swelling and to cleanse and suture the wound afterwards.

## DEHYDRATION

The symptoms of dehydration are sunken eyes and remnants of unsuccessfully shed skin still attached to the body. In mild cases improve your pet's vivarium conditions. Mist spray and ensure that the animal drinks. Drying out (desiccation) can be avoided by offering a humidity chamber or raising the general humidity in the vivarium. Ensure that water bowls are sunk into the substrate to ground level to allow ease of access. Bad cases need veterinarian treatment.

**▲ A simple humidity chamber can be made from an ice-cream container filled with moistened moss or vermiculite to create a humid micro-climate.**

## SLOUGHING

Maintaining the right levels of humidity is one of the key factors in preventing problems when sloughing (ecdysis). The inability to remove old skin properly is known as dysedysis and this can result in a restricted blood supply, especially to blood vessels around toes and fingers.
A humidity chamber – a box containing plastic plants which are well sprayed – will help to reduce this problem in many species and will provide objects for your pet to rub against to aid the slough. Snakes do not eat their shed skin, but many lizards and amphibians will consume this nutritious material.

Common problems include: a spectacle remaining on a snake – this can be removed carefully with moisture and tweezers by an experienced herpetologist; tattered sloughing in a snake or lizard due to inadequate humidity – to remedy this, place your pet into a humidity chamber and/or remove the remains of the skin carefully by hand.

## EXTERNAL PARASITES

Mites and ticks are common problems for reptiles. Mites are little pinhead-like objects that run about a reptile's scales while ticks burrow themselves underneath the scales, especially near the eyes, nostrils or vent. If you maintain good hygiene practices, you are unlikely to have a problem. Mites and ticks are controlled by using Vapona (c). Disinfect furnishings, wash the reptile and place a Vapona block in the vivarium for 24 hours, once a week, for three weeks. This should restrict if not eliminate the problem entirely. However, do not use Vapona near food insects or other pet insects!

## BLISTERS

Especially common on Garter Snakes and other species inhabiting moist habitats. Blisters are often caused by the larvae of nematodes (worm-like parasites). Veterinary treatment is recommended using antiseptic or antibiotics.

## INGESTED SUBSTRATE

This causes major problems for herptiles. Provide food in dishes or bowls and try to ensure feeding animals do not take in this type of material when eating. Keeping an eye on your pet at meal times should prevent this type of problem.

## EUTHANASIA

A terminally sick or aged specimen may need euthanasia. An overdose of barbituates is the most humane method and this should only be administered by a vet.

## AMPHIBIANS AND BACTERIAL INFECTION

Bacterial infections are most common in amphibians because of their delicate skin and aquatic lifestyles. Dead foodstuffs and faecal matter should be removed regularly. Water quality is essential to maintain long-term health in captivity – water should be partly changed regularly. External filtration is always recommended and is probably the single most effective tool in preventing infections in aquatic and semi-aquatic animals. Numerous products are available in the pet trade to filter or condition water for aquatic species. Ensure that any product you purchase is suitable for the species you intend to use it on. All equipment and aquaria require thorough and regular cleaning to minimize problems of this nature.

**◀ This Bearded Dragon has a malformed tail as a result of Metabolic Bone Disease (p. 119). The condition could have been prevented by correct dietary supplementation during growth.**

▲ **This treefrog has lost an eye but it is still able to lead an active life with this disability.**

## SALMONELLOSIS

This disease can be transmitted from many pet animals to man (zoonosis) and vice versa. Salmonella is a genus of bacteria containing over 1500 species, many of which can cause food poisoning. In humans the symptoms are nausea, vomiting and diarrhoea, and in severe cases it may even cause death.

Most salmonella infections in the UK are a result of poor hygiene. Raw chicken is a commonly-infected meat and raw egg is another potential offender.

Many pet animals such as hamsters, rabbits and more exotic species, can be symptom-less carriers of these bacteria, with no infection ever occurring. Any animal with symptoms of diarrhoea, nausea and sudden weight loss should be examined. Infected animals should be put down. The source of infection needs to be identified and eliminated; the culprits are usually old food or bedding.

Do not worry unnecessarily about salmonella. Providing you heed normal animal/human hygiene procedures – washing hands with a micro-bactericidal solution, never cleaning animal housing or equipment in the same area that is used for food preparation, always disinfecting equipment and housing on a regular basis – you have little to worry about.

Follow basic hygiene rules. Keep animals off furniture and out of kitchens. Clean the housing and furnishings on a regular basis with a specially-formulated disinfectant. After contact with any animal, wash your hands. Ordinary soap is a poor cleanser so use a bactericidal hand wash, available from all good chemists.

# Glossary

## A

**Aestivation** The act of passing the summer or dry season in a dormant state.
**Amplexus** A mating embrace seen in some amphibians.
**Aquatic** Growing, living or found in water.
**Arachnid** A class of arthropods characterized by simple eyes and four pairs of legs, which includes spiders, scorpions and ticks.
**Arboreal** Living above ground in trees or bushes.
**Autonomy** The ability to drop or shed the tail as a defence mechanism.

## B

**Brackish** Water that is slightly briny or salty.

## C

**Carnivore** Any animal or plant that feeds on animals; includes amphibians and snakes.
**Casque** A helmet-like structure.
**CITES** The Convention on International Trade in Endangered Species.
**Cloaca** The cavity into which the urinary, alimentary and genital ducts open.
**Crepuscular** Predominantly active at dusk or before dawn.

## D

**Desiccate** To dehydrate or dry out.
**Dewlap** Extendible flap of skin seen on some reptiles.
**Diurnal** Active during daylight hours.
**Dysadysis** The inability to perform ecdysis successfully (see Ecdysis).

## E

**Ecdysis** The periodic shedding of the cuticle in insects and other arthropods and the outer skin layer in herptiles.
**Ecosystem** A system that combines interdependence upon and interaction between living organisms and the immediate environment.
**Ectotherm** Any animal that relies on external heat sources to maintain its body temperature.
**Endoskeleton** The internal structure, or skeleton, of a vertebrate animal.
**Exoskeleton** The external protective structure or cuticle of arthropods.

## F

**Faecal pellets** Bodily waste matter discharged through the anus.

## G

**Gonopods** The paired external reproductive organs of some insects and arthropods.
**Gravid** Pregnant.

## H

**Hemipenes** The paired reproductive organ of a male reptile.
**Herbivore** An animal that feeds on plant matter.
**Hermaphrodite** Possessing both male and female reproductive organs.
**Herpetologist** One who studies amphibians and reptiles.
**Herptile** Any reptile or amphibian.
**Hibernate** To pass the winter in a dormant state with a reduced metabolic rate.
**Humidity** The moisture content of air.
**Hygrometer** An instrument used to measure humidity.

# I

**Instar** The developmental stage of an insect that occurs between any two moults.
**Invertebrate** Any animal lacking a backbone.

# M

**Metamorphosis** Rapid transformation of a larva into an adult – i.e. tadpole to frog.

# N

**Neotony** The term used to describe the reproductive ability of amphibians which still possess larval characteristics.
**Nocturnal** Active mainly at night.
**Nymph** The larval stage of some insects.

# O

**Omnivore** Any animal that feeds upon both plants and animals.
**Ootheca** A capsule containing eggs.
**Oviparous** Producing eggs that hatch outside the female's body.
**Ovipositor** The egg-laying organ of many female insects.

# P

**Parotid gland** A toxin producing gland in amphibians.
**Parthenogenesis** A reproductive process whereby the egg can develop without being fertilized.
**Phase** The term used to describe a variation in the normal form or colour of an animal brought about by seasonal or geographic change.
**Prehensile** Adapted for grasping or wrapping around an object.

# S

**Sexually dimorphic** The ability to differentiate between the sexes by external physical features.
**Slough** Any outer covering that is shed, such as the dead outer layer of the skin of a snake.
**Spermatophore** A jelly-like capsule of sperm.
**Spiracles** Respiratory apertures of insects.

# T

**Taxonomy** The science of classification of living things.
**Terrestrial** Any animal or plant living on the land.
**Territorial** A pattern of behaviour used to defend a specific area.
**Thermoregulation** A method used by ectotherms to maintain optimum body temperature by moving between environments of different temperatures.
**Thermostat** An instrument used to maintain a set temperature.

# U

**Ultraviolet (UV) light** The part of the electro-magnetic spectrum that consists of wavelengths shorter than light and longer than X-rays. UVA (320-400nm) and UVB (260-320nm) are essential for many herptiles. UVC (200-260nm) is dangerous to living organisms.

# V

**Vent** The external opening of the urinary and genital systems.
**Vermiculite** Expanded mica used as a bedding or incubating medium.
**Vertebrate** Any animal with a bony skeleton, including all reptiles and amphibians.
**Vivarium** An enclosure for herptiles and invertebrates.

# Useful Information

## CLUBS AND SOCIETIES

There are a number of clubs and societies that you may wish to join. They are an excellent source of information about all aspects of the natural history and maintenance of exotic species and they are one way to contact like-minded individuals and to obtain livestock. There are also many reptile shows, open to the public, which are advertised in journals and club newsletters.

**Amateur Entomological Society**
PO Box 8774
London SW7 5ZG
www.ex.ac.uk/bugclub

**Association for the Study of Reptiles and Amphibians**
P.O. Box 73
Banbury
Oxon OX15 8RE

**British Herpetological Society**
c/o Zoological Society of London
Regent's Park
London NW1 4RY
www.bhs.org

**British Tarantula Society**
Secretary: Angela Hale
www.thebts.co.uk

**International Herpetological Society**
Secretary: Mr K. J. Hingley
22 Busheyfields Road
Russells Hall
Dudley
West Midlands DY1 2LP
www.international-herp-society.co.uk

## MAGAZINES AND WEBSITES

All the following are available from larger newsagents, pet stores or by subscription. Magazines are a great source of up to date information on the care of reptiles, amphibians and invertebrates, but also offer information on events and shows held throughout the country.

**Reptile Care**
www.reptilecare.co.uk
*A bi-monthly magazine available in the UK*

**Reptiles**
www.reptilesmagazine.com
*A monthly American magazine available in the UK*

**Reptilia**
Salvador Mundi 2
08017
Barcelona, Spain
www.reptilia.net
*A bi-monthly, full colour, 80 page magazine available in English, Spanish and German editions.*

**www.britishcheloniagroup.org.uk**
*Society promoting the care of tortoises and turtles.*

**www.cviewmedia.com**
*A UK site with lots of infomation about pet reptiles and amphibians.*

**www.f-b-h.co.uk**
*The Federation of British Herpetologists exists to promote and support the responsible keeping of reptiles and amphibians by individuals in the UK.*

**www.kingsreptileworld.co.uk**
*A London based specialist breeder and dealer – who actually purchased his first pet snake from the author of the book.*

## FURTHER READING

E.N. Arnold and J.A. Burton, *A Field Guide to the Reptiles and Amphibians of Britain and Europe.* HarperCollins, 1996.

Rev. Gregory Bateman, *The Vivarium.* Upcott Gill, 1897.

A. Bellairs and R. Carrington, *The World of Reptiles.* Chatto and Windus, 1966.

May R. Berenbaum, *Bugs in the System.* Addison Wesley, 1995.

Peter H. Beynon and John E Cooper (Eds), *BSAVA Manual of Exotic Pets.* British Small Animal Veterinary Association, 1991.

J.H. Fabre, *The Insect World of J.Henri Fabre.* Beacon Press, Boston, 1991.

G. Ferguson and P. de Vosjoli (Eds) *Care and Breeding Panther, Veiled and Parsons Chameleons.* Advanced Vivarium Systems, USA, 1995.

Chris Mattison, *Care of Reptiles & Amphibians in Captivity.* Blandford Press, 1983.

John M. Mehrtens, *Living Snakes of the World.* Sterling Publishing Co., 1987.

Pierre Pfeffer (Ed.), *Predators and Predation.* Facts on File Inc., New York, 1989.

Hobart M. Smith, *Handbook of Lizards.* Comstock Publishing, 1946.

Phillipe de Vosjoli, *General Care and Maintenance of Leopard Geckos.* Advanced Vivarium Systems, USA, 1990.

Phillipe de Vosjoli, *Lizard Keeper's Handbook.* Advanced Vivarium Systems, USA, 1994.

Elke Zimmermann, *Breeding Terrarium Animals.* TFH, 1986.

# Index

Major entries are indicated by **bold** page numbers

## T



## U



## V



## W



## X



## Z